D1105791

Wireless Network Administration
A Beginner's Guide

About the Author

Wale Soyinka is an IT systems/network/security consultant with many years of experience in the field. He helps businesses and individuals understand, design, deploy, secure, and integrate wireless networks. Wale is the author of the *Linux Administration: A Beginner's Guide*, now in its fifth edition.

About the Technical Editor

Craig Zacker is a writer, editor, and educator whose computing experience began in the days of teletypes and paper tape. After moving from minicomputers to PCs, he worked as a network administrator and desktop support technician while operating a freelance desktop publishing business. After earning a master's degree in English and American Literature from New York University, Craig supported fleets of Windows workstations and was employed as a technical writer, content provider, and webmaster for the online services group of a large software company. Craig has authored or contributed to dozens of books on operating systems, networking topics, and PC hardware. He has also developed educational texts for college courses and online training courses for the web, and he has published articles with top industry publications.

Wireless Network Administration
A Beginner's Guide

WALE **SOYINKA**

New York   Chicago   San Francisco
Lisbon   London   Madrid   Mexico City   Milan
New Delhi   San Juan   Seoul   Singapore   Sydney   Toronto

The **McGraw·Hill** Companies

Cataloging-in-Publication Data is on file with the Library of Congress

McGraw-Hill books are available at special quantity discounts to use as premiums and sales promotions, or for use in corporate training programs. To contact a representative, please e-mail us at bulksales@mcgraw-hill.com.

Wireless Network Administration: A Beginner's Guide

1234567890   DOC DOC   109876543210

ISBN   978-0-07-163921-7
MHID   0-07-163921-7

Sponsoring Editors	**Technical Editor**	**Composition**
Megg Morin	Craig Zacker	Glyph International
Jane K. Brownlow		
	Copy Editor	**Illustration**
Editorial Supervisor	Lisa Theobald	Glyph International
Janet Walden		
	Proofreader	**Art Director, Cover**
Project Manager	Constance Blazewicz	Jeff Weeks
Vipra Fauzdar,		
Glyph International	**Indexer**	**Cover Designer**
	Claire Splan	Jeff Weeks
Acquisitions Coordinator		
Joya Anthony	**Production Supervisor**	
	Jean Bodeaux	

Thank you Daniel A.

At a Glance

Contents

Part I

Overview

Part II

Hardware

Part III

WLAN, WWAN, WMAN, and WPAN

Part IV

Protocols, Services, and Security in Wireless Networks

Acknowledgments

To the entire McGraw-Hill production group—thank you for your patience with my numerous missed deadlines and tardinesses. "We" did it together.

Introduction

Wireless technologies are all around us. These technologies have pervaded our homes, businesses, and lives. Life without some of these technologies is almost unimaginable.

The inner workings of wireless technologies is one giant puzzle with numerous pieces, consisting of the organizations and individuals who dream up and conceptualize all the wonderful things that can be done wirelessly; the companies that actualize these concepts and continually churn out cool new products; the institutions that help to keep the wireless playing fields fair and safe; and the end users and consumers of these technologies.

And as if this puzzle wasn't complicated enough, individuals and companies have come to expect (and demand) that all these components somehow "auto-magically" connect and communicate with each other as well as with existing wired devices. This last bit is where the wireless network administrator comes in.

Wireless Network Administration: A Beginner's Guide provides a bird's eye view of the various components and technologies with which any wireless network administrator needs to be familiar. The approach in this book is a little different from that of other wireless network administration texts for several reasons: This book does not attempt to prepare the reader for a wireless networking certification exam, it does not focus solely on wireless local area networks (WLANs), and it does not contain complex radio frequency formulae and calculations. Instead, it has a little something for everybody and mostly provides practical and real-world information for network and system administrators who are tasked with understanding, designing, configuring, integrating, supporting, and managing wireless networks.

The book contains five parts and concludes with an appendix on troubleshooting. A mind map of the author's vision and layout of the book is supplied on the publisher's web site at www.mhprofessional.com/computingdownload.

Part I: Overview

This part of the book provides an introduction to the world of wireless technologies. This world is governed mostly by standards (guidelines) that, among other things, aid in ensuring interoperability among similar and countless wireless devices.

To help develop these standards, wireless technical organizations and wireless regulatory organizations develop and help oversee and regulate the use of the common and shared resources over which wireless communications take place. Regulatory organizations play a key role in ensuring that everybody gets a fair and controlled use of the air waves.

Part I also discusses the building blocks of all wireless communications. The concepts and terms surrounding these building blocks are used in virtually all discussions regarding anything wireless, so it is important that you understand these concepts early on in the book.

Part II: Hardware

The standards discussed in Part I are implemented partly in the hardware. And the regulatory bodies discussed in Part I influence how wireless networking hardware transmits and receives wireless signals.

Part II provides an overview of hardware used in wireless networking. Specifically, wireless hardware can be broadly grouped into client- and infrastructure-side hardware. The categorization is based on the function and role of the hardware in the wireless network.

Some wireless hardware is a little more difficult to categorize because it can be found in both client- and infrastructure-side wireless hardware.

Part III: WLAN, WWAN, WMAN, and WPAN

This part delves into the numerous methods used for carrying out wireless communications: wireless local area networks (WLANs), wireless wide area networks (WWANs), wireless metropolitan area networks (WMANs), and wireless personal area networks (WPANs). Which method is best is determined by the specific application or scenario.

The influence of wireless standards is again apparent in the different wireless networking technologies, and some of these standards are examined in greater detail as popular technologies such as WLAN (IEEE 802.11), WWAN (GSM, LTE, UMTS), WMAN (WiMAX), and WPAN (Bluetooth, ZigBee) are discussed.

Part IV: Protocols, Services, and Security in Wireless Networks

Part IV covers some protocols and services that are layered on top of the different wireless technologies. These protocols and services work steadily in the background of most wireless networks to make the network useful to the end user. They serve as the glue for the networks.

Specifically, the Transmission Control Protocol/Internet Protocol (TCP/IP), the Domain Name System (DNS), the Dynamic Host Configuration Protocol (DHCP), authentication services, and proxy servers are covered in Part IV. The wireless network administrator should have a good understanding of the roles, functions, and configurations of these network elements, because they can often make or break the entire network.

Communications over the air waves are known to be notoriously difficult to protect. This is why after all the entities (standards, regulations, technologies, hardware, services, and protocols) are put in place to facilitate communications wirelessly, the next important consideration is to make sure that the communications are secure. Part IV covers the fundamentals of securing wireless networks.

Part V: Wireless Devices Configuration and Other Wireless Network Considerations

This part brings together everything discussed thus far into the practical world. It begins by laying the foundation for a heterogeneous wireless network consisting of different infrastructure hardware and different wireless clients.

These wireless infrastructure devices include wireless access points and an enterprise grade wireless network controller/switches.

The wireless client devices include clients running the mainstream operating systems such as Microsoft Windows, Apple OS X, and FOSS/Linux-based systems. The nuances of the wireless networking implementations (or stacks) on the different platforms are also covered.

You'll also learn how to configure all the devices in a sample wireless network.

Part V ends with a discussion of the process and importance of properly planning, designing, and surveying the network environment before attempting to deploy a wireless network.

Appendix

The appendix provides a quick reference for the wireless network administrator when troubleshooting connectivity issues in a wireless network.

Your feedback is highly welcome, as future editions of this book will build on and grow as a result of your ideas and comments. Comments can be sent to feedback@labmanual.org.

PART I | Overview

CHAPTER 1 | Regulatory and Technical Organizations

Key Skills and Concepts

- Understand the functions and necessity of wireless regulatory organizations.
- Identify some regulatory organizations in the wireless industry.
- Understand the functions and importance of technical organizations in the wireless industry.
- Identify some technical organizations that impact the wireless world.
- Understand the standards creation process.

The wireless technology industry is a multi-billion-dollar industry. But you knew that already.

Oddly enough, a significant factor that drives the huge growth of the wireless industry is the consumer. I say "oddly enough" because in other technology sectors, growth and development tend to be driven by special interest groups, hardware manufacturers, software vendors, trade groups, standards bodies, regulatory organizations, and corporate marketing machines. Instead of the usual technology industry trend, in which the supply drives the demand, wireless consumers' seemingly insatiable appetites for everything wireless are driving the industry to push the technology and discover new frontiers. Consumers demand and expect the latest and greatest in wireless technologies, and they want it now.

A popular expression that states, "With great powers come great responsibilities" can be repurposed by saying this: "With every great wireless technology comes increased bandwidth usage as well as great responsibilities."

Seriously speaking, the increased proliferation of, demand for, and dependence on wireless communication systems have brought issues to the forefront that may not have been such a big deal in the past, such as better security for wireless communications, vastly increased radio bandwidth usage, possibility of abuse of a shared common resource (the airwaves), and better interoperability among wireless devices. In addition, the current balance in the wireless industry wherein the industry must constantly scramble to develop better and faster products to please the consumers as well as fend off their competition is a good recipe for disaster in our wireless technology–dependent world.

Enter the regulatory and technical organizations. This chapter discusses the organizations that help manage and mediate what might otherwise be a very chaotic wireless world. Some of these organizations control what the consumers and wireless industry backers do with the airwaves, and others oversee the technical aspects of how things are done over the airwaves.

Regulatory Organizations

The airwaves over which radio communications travel is a shared natural resource. And because it is a shared and freely occurring medium, people, countries, and industries can choose how they make use of this resource.

NOTE   A medium is the material through which things propagate. Air and space are types of mediums. Radio waves are propagated using air or space as the medium. Airwaves also refer to an intangible medium through which radio signals can propagate.

Numerous entities exist to help enforce or encourage the sensible use of this common resource. These entities function at international, national, and regional levels. Regardless of the level at which they operate, the entities can be controlled by governments, nongovernmental organizations, and profit and non-profit organizations.

Most countries have regulatory and governing bodies that manage the allocation and use of the radio spectrum within the individual country or region. An important function of these bodies is to help reduce or prevent interference in radio communications.

The individual regulatory bodies within each country divide the radio spectrum into two broad groups: the licensed segment and the unlicensed segment. Only users who are at least 18 years old can operate a wireless device in the licensed segment. The operation of wireless devices in the licensed segment of the radio spectrum requires a special permission (license). This license or privilege often comes at a cost. Those younger than 18 years can operate devices using the unlicensed segment of the radio spectrum, in which certain devices can be operated without requiring any special licenses (permissions). Even though the segment is unlicensed, devices and users must still adhere to certain rules that govern the use of this segment of the radio spectrum.

At an international level, bodies such as the International Telecommunication Union Radiocommunication Sector (ITU-R) govern radio frequency (RF) usage, and at the national or regional level, other bodies govern RF usage for the individual nations and/or regions. Table 1-1 shows some examples of regulatory bodies that operate at national/regional levels. The map of the world in Figure 1-1 shows the same information. Note that the list is not all-encompassing; every country not listed here has its own dedicated telecommunication agency whose role and function is equally important to maintaining law and order in radio spectrum usage.

NOTE   The regulatory authorities of each country are often empowered by some kind of national law and/or telecommunications act. The telecommunication act is reviewed and updated as the telecommunications landscape of the country changes over time.

In most countries, the intentions and design of the telecommunication laws are such that they place the citizenry's best interests over that of the commercial entities. But this intention is also subject to change as the telecommunications needs and landscape changes.

Regulatory Body	Country
Comisión Nacional De Comunicaciones (CNC)	Argentina
Australian Communications and Media Authority (ACMA)	Australia
Telecom Regulatory Board (TRB)	Afghanistan
Canadian Radio-television and Telecommunications Commission (CRTC)	Canada
Autorité de Regulation des Communications Electroniques et des Postes (ARCEP)	France
Telecom Regulatory Authority of India (TRAI)	India
Office of Communications (Ofcom)	United Kingdom
Autorità per le Garanzie nelle Comunicazioni (AGCOM)	Italy
Secretaría de Comunicaciones y Transportes (SCT)	Mexico
Infocomm Development Authority of Singapore (IDA)	Singapore
Independent Communications Authority of South Africa (ICASA)	South Africa
Telecommunication Regulation Authority (TRA)	United Arab Emirates
Nigerian Communications Commission (NCC)	Nigeria
Bundesnetzagentur (BNA)	Germany
Federal Communications Commission (FCC)	United States

Table 1-1.   National Radio Communications Regulatory Bodies

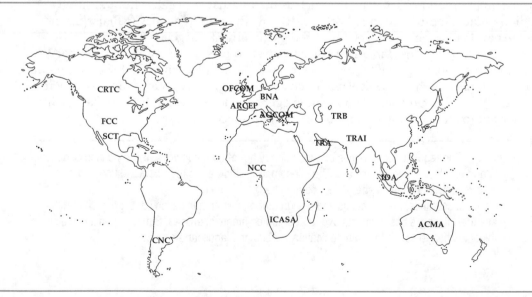

Figure 1-1.   Map of international regulatory bodies

The ITU-R

The United Nations and the ITU-R have at least a couple things in common: they both have established very lofty goals within their individual focus areas and both bodies' decisions can affect the entire world.

What the UN does for the world (facilitate cooperation in international law, international security, economic development, social progress, human rights, achieving world peace, and other good stuff), the ITU-R does for radio communications (help the world avoid total anarchy and chaos that can result from the unregulated, unmanaged, and uncoordinated use of the world's radio spectrum).

Part of the ITU-R's mission statement from its web site states this: "ITU-R is mandated by its constitution to allocate spectrum and register frequency assignments, orbital positions and other parameters of satellites, in order to avoid harmful interference between radio stations of different countries." (The complete text of ITU-R's mission statement can be found at www.itu.int/net/about/itu-r.aspx.)

ITU-R's oversight of the world's radio spectrum use and satellite orbit matters affects the operation of global services such as space research, global positioning systems (GPS), environmental monitoring, mobile communications, and other wireless broadcasting services.

Because the world is such big place, the ITU-R has divided the world into five administrative regions:

- **Region A**   The Americas
- **Region B**   Western Europe
- **Region C**   Eastern Europe and Northern Asia
- **Region D**   Africa
- **Region E**   E-Asia (East Asia) and Australasia

The ITU-R also categorizes the world into three radio regulatory regions, listed here and depicted in Figure 1-2:

- **Region 1**   Europe, Middle East, Africa, the former Soviet Union including Siberia, and Mongolia
- **Region 2**   North and South America and Pacific
- **Region 3**   Asia, Australia, and the Pacific Rim

The regulatory regions shown in Figure 1-2 are actually quite important to the functions of a wireless network administrator. Wireless hardware manufacturers often clearly state (in the product literature or product packaging) the regions in which their products are designed to be operated. Buying and using wireless radio equipment designed for use in one region (such as Region 3) and using the same equipment in another region (such as Region 1) may constitute a punishable offense. In addition to the legal ramifications, using radio equipment outside its designated region may also cause technical interference with other wireless equipment and can disrupt communications.

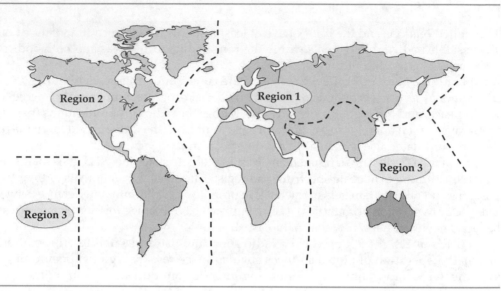

Figure 1-2.   The ITU-R's three radio regulatory regions

Federal Communications Commission (FCC)

The FCC set down the RF communication usage rules in the United States. This government agency is responsible for regulating interstate and international communications by radio, television, wire, satellite, and cable. It is responsible for regulating the use of the radio spectrum in the United States and is empowered by the Communications Act of 1934, which was updated in 1996 and is now referred to as the Telecommunications Act.

The FCC has many rules, including some that affect the operation of wireless devices. For example, a specific FCC rule allows the operation of certain wireless network devices without a license in designated parts of the radio spectrum. This rule can be found in a section of the FCC rules and regulation known as "Title 47 Code of Federal Regulations (CFR) Part 15." It affects the operation of devices used in wireless local area networks (WLANs), cordless phones, devices used in personal area networks (PANs), to name a few. For example, the rule affects the specific frequency ranges in which the devices can operate and the maximum amount of radio energy that the devices can emit. The FCC is there to make sure that operators of wireless radio equipment operate their equipment within the set FCC guidelines.

The FCC also coordinates its actions with those of other international regulatory bodies such as the ITU-R regarding issues that might involve the use of the radio spectrum of other countries. Consider satellite communications, for example.

The FCC acts on behalf of any local wireless operators that want to communicate over the international airwaves. It does this by interfacing with the ITU-R and coordinating the application process with ITU-R.

Office of Communications (Ofcom)

Ofcom establishes the radio spectrum usage rules in the United Kingdom (U.K.). It is concerned with a wide spectrum of communications functions in the U.K.—from managing and allocating the use of the wireless airwaves to making sure that the people in the U.K. get the best from their communication services.

Ofcom is tasked with regulating the following communication sectors: TV and radio broadcasts, fixed line telecoms, cellular (or mobiles), and the airwaves. Its regulations and rules affect the operation of devices used in WLANs, cordless phones, and PAN devices, among others. Ofcom is empowered by the Communications Act of 2003.

It is an independent regulator of communication matters—independent in the sense that it is independent of the nation's government. Ofcom coordinates with other international regulatory bodies such as the ITU-R when it comes to issues that might involve the use of the radio spectrum of other countries, such as in satellite communications. Ofcom represents the U.K. government when dealing with ITU-R.

More complete information can be obtained about Ofcom from its web site at www.ofcom.org.uk/. The organization is people-focused, which is apparent in its easily comprehensible web site.

Australian Communications and Media Authority (ACMA)

ACMA is the de facto body for all matters regarding the regulation of broadcasting, radio communications, telecommunications, and the Internet in Australia. ACMA is responsible for planning and managing the nation's radio spectrum usage and handles all matters regarding compliance with licensing requirements. It investigates complaints of interference with wireless services.

As with other regional regulatory organizations discussed thus far, ACMA also handles all the international radio communications activities. It serves as Australia's representative within the ITU-R.

More detailed information about ACMA's role can be found at its web site, www.acma.gov.au.

Telecom Regulatory Authority of India (TRAI)

TRAI is a government-independent body that is responsible for the regulation of all telecommunications matters in India. Among its other functions, TRAI ensures that India's radio spectrum is used as efficiently as possible. It is responsible for granting licenses to wireless operators, it specifies the terms and conditions of such licenses, and it ensures that the licensee complies with the conditions of the license.

TRAI's mission statement, according to its web site, is "to ensure that the interests of consumers are protected and at the same time to nurture conditions for growth of telecommunications, broadcasting and cable services in a manner and at a pace which will enable India to play a leading role in the emerging global information society."

For more information, see TRAI's web site at www.trai.gov.in.

Canadian Radio-television and Telecommunications Commission (CRTC)

In addition to having the coolest looking logo, CRTC is in charge of regulating all broadcasting and telecommunications activities in Canada. CRTC departs a little from some of the other regulatory bodies in other countries, because its role is more administrative than technical. It defers and reports to another government institution, Industry Canada, for certain technical matters, such as allocating radio frequencies, radio frequency spectrum management, and radio interference.

Another interesting aspect of the CRTC is its hands-off approach to managing the telecommunication sector in Canada. It allows the natural market forces to drive the telecommunications market, but it uses its regulatory powers only where the market doesn't meet the objectives of the nation's Telecommunications Act.

For more information, see CRTC's web site at www.crtc.gc.ca or Industry Canada's web site at www.ic.gc.ca.

Technical Organizations

The focus and objectives of the technical organizations are slightly different from those of the regulatory bodies. These technical organizations are concerned with the inner workings and interoperability of different technologies. They define the problems or needs and come up with solutions to address these issues.

Some of these organizations are responsible for the development and production of numerous standards such as this one, paraphrased from the International Standards Organization/International Electrotechnical Commission (ISO/IEC) Guide 2:2004, definition 3.2: A standard is a document, established by consensus and approved by a recognized body, that provides, for common and repeated use, rules, guidelines, or characteristics for activities or their results, aimed at the achievement of the optimum degree of order in a given context.

A standard is a set of rules that ensures quality and can be in the form of a technical specification. Standards can be developed for processes, protocols, services, and products, and are often regarded as guidelines; as such, compliance to standards is not mandatory. Technical organizations that wish to be regarded as being neutral should definitely not be in the business of enforcing compliance and adherence to the standards they create.

Institute of Electrical and Electronics Engineers (IEEE)

IEEE (pronounced "i-triple-e") is a non-profit professional organization dedicated to the advancement of technology. The IEEE specified standards that conform to the guidelines established by the ITU-R, FCC, Ofcom, and other similar bodies.

The standards formulating process of the IEEE can be a bit tedious and lengthy:

1. A project sponsor submits a Project Authorization Request (PAR) that can be any one of the following:

    ■ A new standard (such as the IEEE 802.11 standard)

    ■ A revision of an existing standard (such as 802.11n, which is a revision of the 802.11 standard)

    ■ Amendments and corrigenda to an existing standard

NOTE   An amendment is a document that contains new material to be incorporated into an existing standard and that may contain technical corrections to that standard as well. A corrigendum is a document that contains only technical corrections to an existing IEEE standard.

2. Once a PAR is approved, future or ongoing work for the proposed IEEE document is officially sanctioned. A working group (or technical committee) is assigned to prepare and develop the document (standard). The working group consists of individuals or organizations affected by, or interested in, the standard.

3. Next comes the draft writing phase. The draft contains the scope, purpose, and outline of the standard. And it is based on the PAR.

4. Ballot stage: A project or draft is ready for a sponsor ballot when it has completed its working group development. This balloting or voting stage goes on until all comments are resolved and a favorable majority of acceptance is reached. If a favorable majority cannot be reached, the standard goes back to the draft stage.

5. Approval or ratification stage: Approval of an IEEE standard is achieved by submitting the document and supporting material to the IEEE-SA Standards Board Standards Review Committee (RevCom), which issues a recommendation to the IEEE-SA Standards Board.

6. Publication stage: Like any good book or document, the approved standard gets a good going over by a professional IEEE standards editor before publication. The editor ensures that the standard is grammatically and syntactically correct. The editor does not make any changes that affect the technical meaning of the standard.

7. Reaffirmation stage: This is a continually ongoing process. Standards typically are valid for a period of five years from the date of approval. During this five-year period, amendments may need to be developed that offer revisions to the original standard. And as a result, before the five-year period elapses, the standard's sponsor must initiate the reaffirmation process that affirms that the technical content of the standard is still valid and the document is reaffirmed for another five-year period. Updates to standards are known as amendments.

NOTE   Amendments are created and standards are updated by task groups. Both the task group and its finished document are denoted an IEEE number followed by a lowercase letter—for example, IEEE 802.11n.

I include this outline of the intricacies of standards and the standards creation process for a couple of reasons:

- The results/outcome of these standards greatly affects us as the end users or administrator who will manage the devices that are built around these standards. These standards directly affect the products that we see on the store shelves.

- A good understanding of standards can help the wireless network administrator make informed and timely decisions about the products used on his/her network. The timing benefits come into play, for example, when a wireless network administrator can properly gauge how long to wait before procuring new wireless hardware that purports to support some future unfinished standard.

Internet Engineering Task Force (IETF)

The IETF is an open international community of technical individuals. Membership in IETF is open to any interested individuals. IETF's stated mission is "to make the Internet work better by producing high quality, relevant technical documents that influence the way people design, use and manage the Internet."

We can infer a few things about this organization from its mission statement:

- The IETF is focused on Internet-related matters and technologies.

- The IETF has a great passion for creating and numbering documents. These documents are referred to as Requests For Comment (RFCs), and there are thousands upon thousands of them.

- The documents that the IETF creates are designed to be used by people who build technologies or products that are used on the Internet.

- The IETF does not refer to itself as a "formal standards organization." Instead, it coyly positions itself as a body of knowledge of which "other" standard bodies or individuals can make use in their technical processes for Internet-related matters and technologies. It has no formal or legal authority to enforce the standards that may result from its work.

NOTE   The IETF's passion for creating and numbering documents is so deep that even its mission statement is documented in one such document: RFC 3935. This document can be found at www .ietf.org/rfc/rfc3935.txt.

Thanks to the volunteer work of the individuals within the IETF, standards such as Transmission Control Protocol/Internet Protocol (TCP/IP) and Hypertext Transport

Protocol (HTTP) have been developed. The fine work of IETF also helps indirectly influence and promote interoperability among different Internet technology implementations.

So what, exactly, has the IETF contributed to the wireless technology world, the impatient reader may ask? The answer is twofold:

- First, the importance of the IEEE in the wireless communications world has already been established. And later on in the book, you will see some of the important wireless standards that IEEE has created.

- Second, the groups at the IETF cooperate with the groups at IEEE in several areas. As a case in point, the effect of work done within IETF can be seen deep within some of the standards that emerge from IEEE. IETF work is used extensively within IEEE wireless standards. Examples of such cooperation can be found in the revised Extensible Authentication Protocol (EAP) Specification (RFC 3748), the EAP state machine specification (RFC 4137), the IEEE 802.1x Radius usage guidelines (RFC 3580), and the Mobile IPv6 Fast Handovers for 802.11 Networks specification (RFC 4260), just to name a few.

True to form, the IETF has created a document clarifying its relationship with the IEEE 802 committee. This document is RFC 4441 and can be found at www.ietf.org/rfc/rfc4441.txt.

European Telecommunications Standards Institute (ETSI)

ETSI is a European standards organization and is officially recognized and sanctioned by the members of the European Union (EU). ETSI is a non-profit body that produces global standards for various information and communication technologies, such as broadcasting, medical electronics, telecommunications, and intelligent transport systems (ITS).

At a global level, ETSI has partnerships and collaborates with other international organizations, such as the ISO, the ITU, and the IEC. Like all good technical standards bodies, these collaborations are meant to help create harmony between ETSI's standards and the standards produced by other organizations. The partnerships also help to reduce or avoid duplication of work done by others.

ETSI been directly involved with popular standards and other technologies, such as Digital Enhanced Cordless Telecommunications (DECT), Global System for Mobile communication (GSM), Enhanced Data Rates for Global Evolution (EDGE), Terrestrial Trunked Radio (TETRA), eHEALTH (healthcare informatics), and ITS, to name a few.

Wi-Fi Alliance

The Wi-Fi Alliance is a non-profit organization comprising several companies devoted to promoting Wi-Fi technology across the globe. The Wi-Fi Alliance can be considered a special interest group (SIG) with a goal of driving the adoption of a single worldwide standard for high-speed wireless local area networking (LAN).

The focus and mission statement of the Wi-Fi Alliance is quite different from that of the IEEE. The Wi-Fi Alliance is not exactly a standards creating entity, but it does try its best to encourage and promote standards compliance among wireless products made by its members. The Alliance tries to promote interoperability and quality among the Wi-Fi products its various member companies supply. To this end, it conducts product certifications, and any product that passes its certifications is allowed to advertise this fact.

The Wi-Fi Alliance has an ever-growing portfolio of registered trademarks, such as Wi-Fi; Wi-Fi Multimedia (WMM); Wi-Fi Protected Access (WPA, WPA2); the Wi-Fi CERTIFIED, Wi-Fi, and Wi-Fi ZONE logos; Wi-Fi Protected Setup. These trademarks are intended to help boost consumers' confidence by assuring them that any products they buy with these logos (trademarks) will be interoperable with other products bearing the same logos.

Summary

This chapter talked about the drivers and the odd dynamics of the wireless technology industry—the odd dynamics being that a lot of the advancements in the industry are driven by wireless consumers' insatiable appetites. Consumers are forcing the suppliers of such technologies to create faster and better products. This is a good recipe for an industry that can quickly become unmanageable in the absence of organizations that can be entrusted with certain oversight functions.

Several regulatory bodies help to regulate the use and allocation of the wireless radio spectrum around the world (among other things). Some of the organizations operate at an international level, some at a regional level, and some at a national level.

Several organizations are influential in developing the standards that drive a lot of wireless technologies, and these standards will be discussed throughout this book. You learned about the structures within some of the organizations, the checks and balances within the organizations, and the standards-making process. Some organizations have a purely non-profit motive for being in the standards game, and some non-profit organizations participate and influence the standards development process.

CHAPTER 2 | Wireless Communication Building Blocks

Key Skills and Concepts

■ Learn the core concepts and components of wireless systems.

■ Understand the properties of the components.

■ Learn about modulation and demodulation techniques.

■ Learn about analog signals and digital signals.

Your basic understanding of the building blocks upon which radio or wireless technologies are built is useful if you plan to have any in-depth discussion on wireless network administration. Many of the important concepts and terms discussed in this chapter build upon one another and are dependent on one another. Our coverage of the concepts will be kept simple and high-level for the most part. Complex mathematical radio frequency (RF) formulae are absent.

Waves

A *wave* is a difficult thing to define but a simple phenomena to understand. It is difficult because a wave can be defined in different ways, based on the specific application or scenario. Waves are easy to understand because you can see their effects in everyday life without having too much technical knowledge.

A wave is a type of disturbance that travels through a medium in a given time; a transfer of energy is always associated with this travel. The medium is a tangible or intangible thing—for example, water, air, wood, wires, or even a vacuum. Each medium has a unique effect on how waves travel. The disturbance created by the wave conveys useful information—the signal or data.

We can refine this definition of a wave to fit the specific focus of this book—wireless communications. Wireless communications are conducted over *radio waves*. We can therefore say that a radio wave is a disturbance that travels through air or space in a given time.

Waves also come in different types: radio waves, sine waves, optical waves, acoustic waves, and electromagnetic waves, to mention a few.

Electromagnetic waves are especially important in wireless communications. They comprise electrical and magnetic components. These wave types are further classified into different subtypes based on certain characteristics of the wave. Types of electromagnetic waves include microwaves, infrared radiation, ultraviolet radiation, X-rays, and finally radio waves.

In wireless communications, information is transferred by systematically changing a characteristic of the radiated waves. Regardless of the application or context, all waves (including radio waves) possess certain characteristics or attributes. These attributes include amplitude, frequency, wavelength, and phase. To convey different meanings (information or data), we simply manipulate some of these wave attributes.

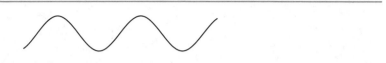

Figure 2-1.   A simple but beautiful sine wave

A *sine wave*, for example, is a mathematical relationship (function) that can be used to describe smooth and repetitive movement. This movement is known as *oscillation*. Conveniently for us, the attributes—amplitude, frequency, wavelength, and phase—that we care about in radio waves can also be extrapolated unto sine waves. This is why sine waves are so important in radio communications. Figure 2-1 shows a simple sine wave.

These characteristics are discussed in detail in the next section.

Frequency

Frequency is a central and measurable characteristic of a wave. Generically speaking, it's a measure of the number of occurrences of a repeating event per fixed unit of time. The frequency of a wave is measured in hertz (Hz). A *hertz* is a unit of frequency equal to one cycle per second. It is measured according to the number of cycles per second that occur—or the number of completed cycles per second.

Applying our definition of frequency to this unit of measurement, we could say the following

- A frequency of 10 Hz means that an event repeats 10 times per second.

- A frequency of 1000 Hz means that an event occurs 1000 times per second.

- A frequency of 1 million hertz (1,000,000 Hz) means that an event repeats 1 million times per second. To avoid writing so many zeros, this can be abbreviated as 1 MHz (one megahertz).

- A frequency of 1 billion hertz (1,000,000,000 Hz) means that an event repeats 1 billion times per second. This can be abbreviated as 1 GHz (one gigahertz).

- A frequency of 2 billion, 450 million hertz (2,450,000,000 Hz) means that an event repeats 2.45 billion times per second. This can be abbreviated as 2.45 GHz (2.45 gigahertz).

- A frequency of 1 trillion hertz (1,000,000,000,000 Hz) means that an event repeats 1 trillion times per second. This can be abbreviated as THz (one terahertz).

 NOTE   Frequencies in multiples of gigahertz are very important to humans. We use the 2.45 GHz frequency range to cook food in microwave ovens. Coincidentally, this is the same frequency range at which radio waves used in a type of wireless local area network (WLAN) technology operate.

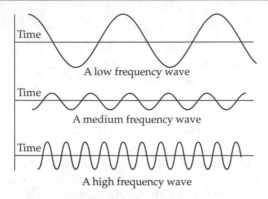

Figure 2-2.   Three sine waves at different frequencies relative to each other

Figure 2-2 compares the frequencies of three waveforms. The topmost sine wave has the lowest frequency, the middle wave has a medium frequency (relative to the other two waves), and the bottom wave has the highest frequency.

 NOTE   Radio frequency (RF) is a specific type of frequency that forms the cornerstone of wireless communication technologies and most of the IEEE 802.11 family of standards. It is a frequency (or rate of oscillation) within the range of 3 Hz to 300 GHz.

Wavelength

Just as in most aspects of real life, a wave consists of successive troughs (lows) and crests (highs). The distance between two adjacent crests or troughs is called the *wavelength* (Figure 2-3). Wavelength can be measured in several ways—one crest to the next crest, or one trough to the next trough.

Wavelength is related to frequency, and the relationship is inversely proportional—in other words, the higher the frequency, the shorter the wavelength, and the lower the frequency, the longer the wavelength.

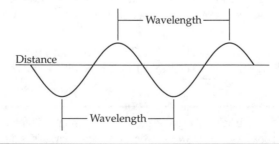

Figure 2-3.   Wavelength

A wavelength affects how the wave will interact with the medium in which it travels. For example, waves with shorter wavelengths are more easily affected by solid objects (walls, building, trees, furniture, and so on) that lie in their path. And waves with longer wavelengths can propagate over longer distances.

Amplitude

Amplitude is a measure of the magnitude (the relative size or extent) of a wave (Figure 2-4). It is a measure of power, strength, or height of a radio wave—the signal strength. Amplitude is formally defined as the maximum displacement of a periodic wave.

Amplitude affects the *juice* remaining in a signal after traveling over a certain distance. An RF signal transmitted with an initially higher amplitude will not necessarily travel farther compared to a signal transmitted with an initially lower amplitude, but the signal with the higher amplitude will be more useful when it arrives at its intended destination.

While the amplitude of a periodic wave can change as it propagates through space, its frequency remains the same. The amplitude change can be a reduction in the signal strength (*attenuation*) or an increase in the signal strength (*amplification*). Various components in wireless devices are responsible for signal attenuation and amplification. For example, wiring and connectors can cause attenuation, while an antenna can cause amplification.

In wireless communication systems, such as WLANs, amplitude can be measured at the transmitting end and at the receiving end of the communicating entities. At the transmitting end, this value is called the *transmit amplitude* and at the receiving end it is called the *received amplitude*. The wireless signal strength indicator in various wireless devices such as cell phones and wireless cards in computers often use the received amplitude to indicate the strength of the wireless signal to the user—to show how far or near the user is from the source of the wireless signal (the *transmitter*).

Phase

This is where things start to get a bit complicated. The *phase* of a wave is the offset of the wave from a reference point; it is a relative measure between two quantities—one

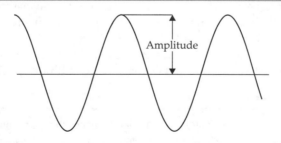

Figure 2-4.   Amplitude of a wave

known and the other unknown—that have something in common. When comparing the phase relationships at a given time between two quantities, the phase of one is usually assumed to be zero, and the phase of the second quantity is described with respect to the first.

Here's an attempt to untangle this convoluted definition of phase: The relationship between the two quantities can be expressed in terms of their degrees of separation (0 to 360 degrees). Therefore, phase can also be described in terms of an angular relationship between two waves. For example, two waves that are *completely* in-phase will be said to be 0 degrees out-of-phase (or, they are simply *not* out-of-phase). And two waves that are *completely* out-of-phase are said to be 180 degrees out-of-phase.

Phase is a subtle concept with profound influence on RF communications. In particular, phase affects the amplitude of a wave, as you will see in the coming sections.

Phase Difference

Any two wave sources (oscillators) are said to have a *phase difference* if they have the same frequency but different phases. The oscillators are then said to be out-of-phase with each other. When that happens, the phase difference determines whether the waves reinforce or weaken each other.

Constructive Interference   When the crest (high) of one wave passes through, or is superimposed upon, the crest (high) of another wave, the waves are said to *constructively interfere* with one another. The converse is also true: constructive interference can occur when the trough (low) of one wave passes through or is positioned upon the trough (low) of another wave.

So, given any two waves that are in-phase, the resulting wave or combined waves will be stronger than either of the individual waves. This simply means a stronger signal at the receiving end, and *stronger* here also means *higher amplitude*.

From a wireless network administrator's perspective, most forms of constructive interferences are beneficial to a wireless network because they can enhance wireless communications.

Destructive Interference   When the crest (high) of one wave passes through, or is positioned upon, the trough (low) of another wave, the waves are said to *destructively interfere* with one another.

One type of destructive interference is the effect of two waves that are out-of-phase with each other, especially when they cancel each other out. This results in a diminished (lower amplitude) signal.

As a wireless network administrator, you almost always want to keep all forms of destructive interference to a minimum, because they can hinder or corrupt wireless communications.

NOTE   During constructive or destructive wave interference, amplitudes add or subtract depending on whether the interaction is between the two highs or two lows or between a low and a high.

Bands

The radio spectrum is divided into different communication frequencies called *bands*. Several models exist for grouping frequencies into bands, such as the following:

- **Medium Wave (MW) band**   This band is used mainly for AM broadcasting. In most parts of the world, the frequency range is 531–1602 kHz.

- **Very High Frequency (VHF) band**   These are the radio frequencies in the range from 30 to 300 MHz. The common FM broadcast band is within this range. In most parts of the world, the FM band is in the 87.5–108 MHz range, but the FM band in Japan is 76–90 MHz.

- **NATO K band**   This band encompasses frequencies between 20 and 40 GHz.

- **IEEE K band**   This band encompasses microwave frequencies ranging from 18 to 27 GHz.

- **Industrial, Scientific, and Medical (ISM) band**   Frequencies in the ISM band are defined by the ITU-R (see Chapter 1). Each country's regulatory body controls how the frequencies are used. Popular WLAN standards such as IEEE 802.11b and IEEE 802.11g operate in the ISM frequency band ranges. Table 2-1 shows the frequency ranges available in the ISM band.

Frequency Range	Center Frequency
6.765–6.795 MHz	6.780 MHz
13.553–13.567 MHz	13.560 MHz
26.957–27.283 MHz	27.120 MHz
40.66–40.70 MHz	40.68 MHz
433.05–434.79 MHz	433.92 MHz
902–928 MHz	915 MHz
2.400–2.500 GHz	2.450 GHz
5.725–5.875 GHz	5.800 GHz
24–24.25 GHz	24.125 GHz
61–61.5 GHz	61.25 GHz
122–123 GHz	122.5 GHz
244–246 GHz	245 GHz

Table 2-1. ISM Frequencies

Name	Frequency Range
UNII Low	5.15–5.25 GHz
UNII Mid	5.25–5.35 GHz
UNII Worldwide	5.47–5.725 GHz
UNII Upper	5.725–5.825 GHz

Table 2-2.   UNII Frequencies

■   **Unlicensed National Information Infrastructure (UNII) band**   Frequencies and power limits in the UNII band are defined by the FCC in the United States. This band is subdivided into frequency groups that are each 100 MHz wide. This band is generally used for high-bandwidth applications (such as Voice over IP, video, and so on) because of its capacity to handle such applications. Higher bandwidth means higher data rates. The popular IEEE 802.11a WLAN standard operates in the UNII frequency band range. See Table 2-2.

Channels

A *channel* in communications parlance is the route through which information is sent. To maintain some semblance of sanity, the RF spectrum bands are divided into channels or groups with fixed widths. Channels are usually set aside within bands for the same purpose.

Referring to radio frequencies in terms of channels is easier than quoting actual radio frequencies. For instance, asking a layperson to tune her television to channel 3 is easier than asking her to tune her television to the 61.25 to 65.75 MHz frequency to receive the video and audio components of a television broadcast.

Table 2-3 shows how channels map to actual frequencies in the ISM band. These frequencies are used in the WLAN communications that are based on the IEEE 802.11b, IEEE 802.11g, and IEEE 802.11n standards.

Modulation

The concept of *modulation* is at the heart of RF communications. It is all about developing efficient ways of massaging a signal so that it can be transmitted and properly interpreted from one place to another over a given medium. Modulation is the process of varying a wave pattern to use that variance to convey some kind of information. A device that performs modulation is known as a *modulator,* and a device that performs the reverse operation of modulation is known as a *demodulator*.

Channel	Lower Frequency (GHz)	Center Frequency (GHz)	Upper Frequency (GHz)
1	2.401	2.412	2.423
2	2.404	2.417	2.428
3	2.411	2.422	2.433
4	2.416	2.427	2.438
5	2.421	2.432	2.443
6	2.426	2.437	2.448
7	2.431	2.442	2.453
8	2.436	2.447	2.458
9	2.441	2.452	2.463
10	2.451	2.457	2.468
11	2.451	2.462	2.473

Table 2-3.   How Channels Map to Frequencies in the ISM Band

Several modulation schemes exist in the communications world, and in wireless communication systems, we don't get to choose any random modulation scheme; instead, we have to choose the technique that best solves the problem at hand or the technique that best suits our application. Depending on the problem or the application, we can choose from spread-spectrum modulation techniques, digital modulation techniques, or analog modulation techniques.

Spread-Spectrum Modulation Techniques

Spread-spectrum modulation techniques are characterized by their signature, wherein the transmitted signal occupies more bandwidth than the actual information being modulated. You could say that in spread-spectrum techniques, the carrier signals occur over the full bandwidth or spectrum of the transmitting frequency. This is akin to a car (the carrier) occupying two lanes on a highway (the medium) by driving in the middle lane, even though all the car actually needs is one lane to transport its passengers (information) to their destination.

Wireless applications that make use of spread-spectrum modulation are also characterized by their ability to function at low power levels, or amplitudes.

FHSS

Frequency-hopping spread spectrum (FHSS) is a legacy spread-spectrum technology used in RF communications. It works by transmitting data using a specific frequency for a given length of time and then changing (hopping) to another frequency to transmit more data for another fixed length of time, and this repeats *ad infinitum* until all the data transfer is completed.

FHSS suffers a drawback when too many FHSS-based devices are operating within the same area, because the probability of such devices hopping to the same frequency is increased.

The popular wireless technology Bluetooth, used in personal area networks (PANs), uses FHSS.

DSSS

Direct-sequence spread spectrum (DSSS) is a modulation type based on spread-spectrum techniques. Spread-spectrum technologies deliberately transmit signals using more frequency bandwidth than is necessary to transmit the intended information. This simply increases the signal-to-noise ratio (SNR). DSSS implementation of spread-spectrum adds some noise signal to the actual signal being transmitted; the extra noise is in the form of random –1 and 1 values. Timing is very important in DSSS; the station doing the transmitting of the data takes care of modulating the signal (adding the extra noise), and the receiving station needs to know how to remove the extra noise from the signal to obtain the original data.

Relative to FHSS modulation techniques, DSSS offers these advantages:

■   Better resistance to interference

■   Better resistance to interception

■   Better multi-access capability (supports multiple users transmitting simultaneously on the same frequency)

■   Provides slightly higher data rates

■   Shorter delays

FHSS-based systems and DSSS-based systems generally do not coexist well—that is, one will likely interfere with the other. The interference can occur in either direction. For example, because DSSS-based devices try to transmit on every frequency in the band, FHSS systems might not be able to find free channels to use. Conversely, the constant hopping between available frequencies in the band by FHSS-based systems may starve DSSS-based systems of frequency resources.

The IEEE 802.11b WLAN standard uses the DSSS form of modulation.

Digital Modulation Techniques

Digital modulation techniques are employed when there is a need to convert digital signals to analog signals. The converse of digital modulation would be demodulation—the conversion of analog signals to digital form.

In the digital modulation world, there are currently three methods used to perform this conversion, and they all depend on varying some attributes of a sine wave representing the signal. Again, these attributes are frequency, amplitude, and phase.

The current methods for performing digital modulation are Frequency-Shift Keying (FSK), Amplitude-Shift Keying (ASK), and Phase-Shift Keying (PSK).

NOTE   In communication systems, the original information or original signal element in the system is called the *baseband*. The baseband is an aspect of the system as it exists before any type of modulation has been applied to it.

Frequency-Shift Keying

Digital-to-analog conversion of signals is achieved in FSK by changing the frequency property of the sine wave representing the signal. As you might expect in all things digital, two values are involved in FSK. The modulation system can choose to represent a particular frequency with a 0 (zero) and the other frequency with a 1 (one). And by stringing the many possible combinations of these 0's and 1's together, we can convey different meanings or information signals.

OFDM   Orthogonal frequency-division multiplexing is a mouthful modulation technique! OFDM is a digital modulation technique—specifically, an implementation of the broader frequency-division multiplexing (FDM) techniques. FDM is a three-part phase—frequency, division, and multiplexing—that involves dividing a frequency into chunks (subcarriers) so that the individual chunks (or divisions) can be transmitted individually over a single communications channel or multiplexed.

OFDM extends the basic FDM techniques by using subcarriers that are *orthogonal* (or uncorrelated) to each other.

OFDM provides more efficient use of the spectrum, better resilience to severe channel conditions, and less sensitivity to time synchronization errors between the sender and receiver compared to in DSSS. Instead, frequency synchronization is a more important factor in OFDM.

OFDM is widely used in numerous applications: for example, the WLAN IEEE 802.11(g)(n)(a) standards, the wireless MAN technology (WiMAX) based on IEEE 802.16, some digital radio systems, and mobile broadband wireless access standards such as IEEE 802.20 and IEEE 802.16e.

Amplitude-Shift Keying

Digital-to-analog conversion of signals is achieved in ASK by changing the amplitude of the wave according to changes in the information signal. The other properties of the sine wave, such as the frequency and phase, are kept constant while the amplitude is varied. As with all things digital, two values are involved in ASK. The modulation system can choose to represent a particular amplitude with a 0 and another amplitude with a 1. By stringing the many possible combinations of these 0's and 1's together, we can convey different meanings or information signals.

Phase-Shift Keying

Digital-to-analog conversion of signals is achieved in PSK by changing the phase attribute of the sine wave representing the original signal. Again, like all things digital, two values

QAM

Hybrid modulation techniques can be created when we mix-and-match any of our standard digital modulation methods such as PSK, ASK, and FSK. One such modulation technique is the result of the marriage of ASK and PSK and is known as *Quadrature Amplitude Modulation (QAM)*. In QAM, the amplitude and phase of the sine wave representing a signal are changed or varied in response to changes in the data or information signal.

are involved in PSK: The modulation system can choose to represent a particular phase with a 0 and another phase with a 1. By stringing the many possible combinations of these 0's and 1's together, we can convey different meanings or information signals.

Analog Modulation Techniques

Analog signals are said to be continuous and long-winded in nature. They can go on forever if uncurtailed. Try not to be too confused when it is said that analog signals can go on *forever* and then in the same breath said that analog signals have *start* and *ending* points. Basically, this means that the possible variations within the signals themselves can be varied and continuous.

The human voice is a good example of an analog quantity. Within any human voice is a measurable upper and lower limit. But within these limits are numerous variations. This is why some people scream loudly and some scream softly, some whisper quietly and some whisper loudly, some talk normally and some talk softly, and so on. We say that some people can change the pitch, volume, timbre, or tone of the sound they produce. Singers especially learn how to control and manage the characteristics of the vocal sounds they produce.

So why should the wireless network administrator care about analog signals or digital signals? It's simply because *all* things wireless are analog by nature. The wireless transmission medium itself is analog.

Computers and other electronics often process and output information in digital formats. If the signal information is to be used and processed only within the computer system, it can remain and die in its digital form. But when it comes to communicating the digital information wirelessly to other entities, the information has to be repackaged into a form that can be transmitted using the analog wireless medium. Forget and ignore all the books, TV shows, news media, and people that say that we are living in a digital world. We are not: Our world is still analog. Nature is analog. The digital world we have created is living inside our analog world.

Can you see where this is going? Enter analog modulation techniques, which are good for transmitting analog signals over analog mediums. Here are some examples of analog modulation methods.

Amplitude Modulation (AM)

This analog modulation technique works by varying the amplitude of the transmitted signal relative to the information signal. A key difference between AM and its distant cousin, ASK, is that AM does not aim to map the changes in amplitude to discrete digital values of 0's and 1's. And, as such, we wouldn't normally use AM as a modulation technique for WLAN, wireless personal area network (WPAN), or wireless wide area network (WWAN) communications that deal with digital and analog signals. But we would use AM as a modulation technique for transmitting the analog human voice signal over the analog wireless medium.

AM is not used in today's bandwidth-hungry wireless communication systems for several reasons. One of these reasons is that the transmitter power usage of AM is very inefficient.

A popular implementation of AM is found in AM radio broadcasts.

Frequency Modulation (FM)

This analog modulation technique works by varying the frequency of the transmitted signal relative to the information signal. A key difference between FM and its distant cousin, FSK, is that FM does not aim to map the changes in frequency to discrete digital values of 0's and 1's. So we wouldn't normally use FM as a modulation technique for WLAN communications, which deal with digital and analog signals. But we would use FM as a modulation technique for transmitting the analog human voice signal over the analog wireless medium.

A popular implementation of FM is found in FM radio broadcasts.

Summary

In this chapter on wireless communication building blocks, you learned the basic premise of waves. And because waves (radio waves) are so important, some characteristics of waves—frequency, amplitude, modulation, phase, channels, and so on—were covered. These characteristics make up an important part of the wireless communications jargon that we as wireless network administrators should be able to understand and speak. You should now understand some of the methods available that rely on tweaking the characteristics of waves to communicate some useful information signal.

This chapter also dispelled a popular myth, that we are living in a digital age. Beyond all reasonable doubts, it was proven that we are actually living in an analog world; the so-called digital world is living inside our analog world.

CHAPTER 3 | Wireless Standards

Key Skills and Concepts

■ Understand the importance and role of standards.

■ Identify some popular and not-so-popular standards that define some common wireless technologies.

Chapter 1 discussed some noteworthy organizations, including those with regulatory functions and some with purely technical functions and objectives. You learned how tedious and involved the standards development process can be within the technical bodies. Surely something good should come from all that hard work—right?

This chapter discusses some of the standards that were borne out of these organizations, focusing on standards that impact wireless technologies in one form or another. Standards can be moving targets that change often. This chapter covers some legacy standards, though not in depth, with a focus on current and future standards.

Standards

As with almost every other aspect of our modern world, regulatory standards regarding wireless use and technologies exist to guide us and act as a reference point for interoperability, efficiency, and other aspects of wireless technologies. Remember that standards are merely guidelines and are not binding, and as such people, countries, and industries can choose whether or not to adhere to them.

Standards are important for several reasons, including the following:

■ **Interoperability**   Standards help to promote interoperability among devices made by different vendors. Individual vendors can build their devices to conform to a particular standard, ensuring that their devices will be able to work together.

■ **Efficiency**   Every vendor can reuse existing solutions that a particular standard has addressed, instead of having to re-create individual solutions. The idea is that a standard is developed once and can be used in many instances with many products.

■ **Prevention of vendor lock-in**   Standards give technology consumers the freedom to choose and buy whatever products they like, which helps to reduce or prevent consumers being locked into solutions or wares provided by particular vendors. When solutions are designed according to standards and specifications, third-party vendors can provide continued support or complementary solutions when an original vendor is no longer able or willing to support its products.

IEEE 802.3

The Institute of Electrical and Electronics Engineers (IEEE) 802.3 standard is a collection of IEEE standards that define the properties as well as the working characteristics of the Physical Layer and Data Link Layer's media access control (MAC) sublayer of *wired* Ethernet.

Why in the world is a standard for wired networks in a book dedicated to wireless networks? And to make matters worse it's right at the top of the list of the other wireless standards that will be discussed later on.

The reason is simple. This is the standard that most wireless standards and wireless technologies want to be like when they grow up.

IEEE 802.3 is the great granddaddy of networking protocols. It has the feel and functionality that most wireless standards want to emulate. It represents the behavior and nirvana that wireless network engineers would like to re-create for wireless devices.

All this is not to say or imply that IEEE 802.3 is a perfect standard or that network engineers consider it perfect. It isn't perfect, but it has been around for so long and has served us so well that most agree it could be used as a basis or starting point for defining new wireless networking standards.

Moreover, until such time when the world becomes completely cable-free (wireless), it will be necessary for wireless devices to interoperate with their wired or cabled siblings. And so it was deemed worthwhile that the wireless standards be designed so that they are similar to the wired standards.

IEEE 802.3 is especially concerned with local area network (LAN) technologies, with a nod to some wide area network (WAN) applications. And this is okay, because the networked world is, after all, like one big LAN—that is, WANs are simply a collection of individual LANs. So now that we have considered this little issue of wired networking standards in a book about wireless, we can move on to the wireless standards proper.

IEEE 802

The IEEE 802 is not a single standard—it refers to a family of standards. The committee within IEEE that is responsible for this group of standards is known as the IEEE 802 committee, and it deals with LAN, metropolitan area network (MAN), and personal area network (PAN) technologies and standards. It is concerned mostly with the data link and physical layers of the Open Systems Interconnect (OSI) model.

NOTE   You can learn more about the work done by IEEE 802, at its web site: www.ieee802.org.

IEEE 802.11

The IEEE 802.11 (pronounced 802 dot 11) standard can be described as the father of all WLAN standards. It evolved from a subcommittee within the IEEE 802 committee, which is implied by the dot notation (the period) used in the standard number.

This standard and its various revisions are the crux of most of our discussion in this book. The IEEE 802.11 standards comprise various individual standards that cover wireless networking technologies. These standards are forever evolving and adapting to meet technology and industry needs. At the time of this writing, the most current complete revision of this standard is IEEE 802.11-2007; however, several amendments have been made to the standard after the 2007 major revision. Table 3-1 shows some of the IEEE 802.11–based standards.

NOTE   It's easy to take for granted the fact that the original IEEE 802.11 standard exists and was actually a standalone standard that served an important purpose when it was first created. To be pedantic, the IEEE 802.11 standard specified data rates of 1 megabit per second (Mbps) and 2 Mbps and operated in the 2.4 gigahertz (GHz) band. All the other IEEE 802.11 standards with letter designations (such as 802.11b) are amendments to this original standard.

Amendments to the 802.11 standards are discussed next. And in continuing with the genealogical analogy, some of these amendments can be viewed as the offspring of the IEEE 802.11 standard.

IEEE 802.11b

The emergence of this standard represents the period when WLAN technology started to be taken very seriously and gained mainstream adoption.

The IEEE 802.11b standard specifies a maximum raw data rate of 11 Mbit/s. In its time, it was a dramatic improvement in the data rate compared with that offered by the preceding IEEE 802.11 standard (11 Mbit/s vs. 2 Mbit/s). It uses the direct sequence spread spectrum (DSSS)–based modulation scheme.

The standard specifies operation in the 2.4 GHz band, which makes WLAN devices subject to interference from a plethora of other everyday devices that operate in the same frequency range (such as microwave ovens and cordless phones).

Standard	Center Frequency	Bandwidth (MHz)	Data Rates (Mbps)	Modulation
802.11a	5 GHz	20	6, 9, 12, 18, 24, 36, 48, 54	OFDM
802.11b	2.4 GHz	20	1, 2, 5.5, 11	DSSS
802.11g	2.4 GHz	20	1, 2, 6, 9, 12, 18, 24, 36, 48, 54	OFDM and DSSS
802.11n	2.4 GHz	20	7.2, 14.4, 21.7, 28.9, 43.3, 57.8, 65, 72.2	OFDM
802.11n	5 GHz	40	15, 30, 45, 60, 90, 120, 135, 150	OFDM

Table 3-1.   IEEE 802.11 Wireless Standards

NOTE   The maximum data rates specified by any of the 802.11 standards are achievable only in a *greenfield*. A greenfield is an engineering jargon that describes an almost perfect environment. For our application, *greenfield* refers to a wireless environment that lacks any constraints (such as interference) imposed by existing wireless networks or other sources of interference.

IEEE 802.11a

The IEEE 802.11a standard specifies operation in the 5 GHz Unlicensed National Information Infrastructure (UNII) frequency bands. It uses the same OFDM (orthogonal frequency division multiplexing) modulation used in IEEE 802.11g (which is discussed next). A maximum data rate of 54 Mbit/s is specified.

Because 802.11a does not operate in the crowded 2.4 GHz band, it is less prone to interference. However, the higher frequency used results in a couple of disadvantages: the overall range achievable is reduced, and the radio signals are more easily absorbed, and therefore lost, due to the smaller wavelength of the 5 GHz radio waves.

Even though IEEE 802.11a–based devices can happily coexist with other wireless devices that function in the 2.4 GHz industrial, scientific, and medical (ISM) bands, the two devices cannot communicate with each other. They don't operate in the same frequencies, and the differences in the modulation techniques used by each also prevent them from communicating. IEEE 802.11a devices use OFDM, while IEEE 802.11b devices use DSSS.

Devices that implement this standard are not as widely available in the marketplace as devices that support the other standards.

IEEE 802.11g

The IEEE 802.11g standard specifies operation in the 2.4 GHz ISM band frequency. The standard specifies a maximum raw data rate of 54 Mbit/s. It uses a variant of the OFDM–based modulation scheme as well as the DSSS modulation technique.

Hardware based on the IEEE 802.11g standard is backward-compatible with IEEE 802.11b–based hardware. This backward-compatibility was a big plus that worked in favor of the adoption of this standard, because users and system integrators were sure that their investments in the early IEEE 802.11b hardware would not go to waste.

Because maintaining backward-compatibility with existing IEEE 802.11b devices was a key principle guiding the design and specifications of the 802.11g standard, it identifies three different modes of operation for the devices:

- **g only mode**   IEEE 802.11b clients are not invited to this party. Devices operating in this mode use only the OFDM modulation scheme and are able to support higher data throughputs than would be possible if they were operating in any of the other modes.

- **b only mode**   Only wireless clients that are prepared to operate using DSSS modulation scheme are invited. This is sort of a downgrade for 802.11g–capable devices, because they are forced to use slower rates for communications. The overall throughput in this mode of operation is roughly equal to that of the slowest devices—in this case, the 802.11b devices—and, as such, data rates of 11 Mbit/s and less should be expected.

- **b and g mode**   Everybody is invited to this party. This includes the devices that support DSSS and/or OFDM modulation schemes, which include the 802.11b and 802.11g devices.

Even though it is an open party, some rules dictate the behavior of the attendees. This rule is referred to as "protection mechanism." It optimizes the performance of the wireless network when both 802.11b and 802.11g clients are present.

Like other wireless hardware that functions in the crowded 2.4 GHz frequency band, IEEE 802.11g devices suffer from interference from other devices that operate in the same frequency band.

IEEE 802.11n

At the time of this writing, IEEE 802.11n is quickly gaining wide acceptance. Most equipment manufacturers include several products in their portfolios that implement the standard in one form or another. The standard specifies operation in the 2.4 and 5 GHz frequency bands. It can offer maximum data rates of up to 600 Mbit/s. It uses the OFDM-based modulation scheme.

IEEE 802.11n offers overall improvements in comparison to the preceding standards. One significant change that facilitates some of these improvements is its use of the Multiple Input/Multiple Output (MIMO) technology. MIMO technology essentially uses multiple antennas for transmitting and receiving, which lets it send and receive more information than the standard dual- or single-antenna setups. *Channel bonding*, which uses two non-overlapping and adjacent channels in the 5 GHz frequency band for radio frequency (RF) communication, is also used in 802.11n. It is also known as 40 MHz, because it combines two channels, each with a width of 20 MHz.

The standard defines several built-in mechanisms that aid in the coexistence of 802.11n, 802.11b, 802.11g, and 802.11a wireless devices. One such mechanism allows 802.11n devices to embed their transmission inside 802.11g or 802.11a transmissions. This, of course, makes the 802.11g and 802.11a devices happy, because it looks like everybody is speaking the same language and can transmit at maximum possible speeds.

The vast improvements in throughput offered by 802.11n make it attractive for use with applications such as streaming video, wireless Voice over IP (VoIP), video conferencing, and others.

NOTE   For better or worse, we can thank the Wi-Fi Alliance for products in the marketplace that implement standards that have not yet been fully ratified by the IEEE. Better because the Alliance bypasses some of the red tape and bureaucracy in the standards-development process and helps consumers get cool and new technologies more quickly. Worse because some of the standards certified by the Alliance are not fully mature and are not ratified.

Consider the Alliance's certification of the IEEE 802.11n Draft 2.0 products long before the standard became ratified by IEEE. Several vendors were able to make and sell products that supported a standard that was still under development. I mention this only for historical reasons, because IEEE 802.11n has, of course, since become a fully ratified standard.

IEEE 802.11y

The IEEE 802.11y standard currently specifies operation of high-powered devices in the 3650 to 3700 MHz RF band in the United States. Its focus and primary application is not for off-the-shelf and casual use. Devices implementing this standard can theoretically link distances as far apart as 3 miles (5 kilometers).

Successful deployment of hardware implementing this standard is dependent on minimal interference from other RF devices. And to this end, the standard specifies several mechanisms that can help.

The first is the requirement of operators to obtain a so-called "lite license" before being granted permission to operate base stations (also referred to as enabling stations) implementing this standard. The license is nonexclusive.

The idea is that operators request and obtain a license from a regulatory body (such as the U.S. Federal Communications Commission). The licensing requirements also involve the base station operator specifying the physical location of high-powered base station(s), which help in connecting the end user client devices that need to communicate with or via the base stations.

The client devices do not need to pay for or request any special licenses to implement this standard. To participate in an 802.11y network, the client devices require an "enabling signal" from the licensed base station.

Another important concept is the Extended Channel Switch Announcement (ECSA), which allows RF devices to dynamically change or select their current RF channel to a channel that has the most favorable noise and quality characteristics.

IEEE 802.11k

This amendment to the IEEE 802.11 standard provides specifications on how to make better use of available radio resources. It also details how better to manage the radio resources by exposing radio and network information to facilitate easier maintenance and management.

The IEEE 802.11k standard also aims to provide a standardization for new mechanisms to supply the required measurements and metrics that will assist in more efficient radio resource management.

It allows wireless clients to make more efficient and intelligent use of the radio resources. For example, it can help a mobile client make a decision on the best access point (AP) to connect to, by considering useful parameters (such as AP utilization) other than signal strength.

IEEE 802.11i

IEEE 802.11i defines security mechanisms for wireless networks.

Unlike most of the other standards discussed thus far, IEEE 802.11i is not intended to be used by itself as a wireless communication protocol. Instead, it is intended to complement and or be layered on top of the other standards that deal with the actual communications parameters.

The original IEEE 802.11 standard specified the use of 802.11 Wired Equivalent Privacy (WEP) as a security mechanism. It was soon discovered, however, that several

WEP weaknesses made it undesirable for use in protecting WLANs. And, true to form, the IEEE 802.11 working group set up a task group to come up with a replacement security solution, which is described in the IEEE 802.11i standard, more commonly known as Wi-Fi Protected Access 2, or WPA2.

NOTE   When the brokenness in WEP was discovered, the Wi-Fi Alliance came up with a stop gap security mechanism to replace WEP: Wi-Fi Protected Access (WPA). Weaknesses have since been discovered in WPA, too.

IEEE 802.15

IEEE 802.15 refers to an IEEE 802 subcommittee that governs standards covering various wireless personal area network (WPAN) technologies.

Several task groups exist within the IEEE 802.15 subcommittee and were formed to focus on different WPAN implementations:

- **Task Group 1**   This group developed standards that were based on some of the early specifications developed by the Bluetooth special interest group. IEEE standards exists for Bluetooth versions 1.1 and 1.2.

  This task group was short-lived and came up with only a couple of IEEE standards before disbanding.

- **Task Group 2**   This task group deals with issues concerning the coexistence of devices that implement WPAN standards and those that implement the WLAN standards.

- **Task Group 3**   This group developed standards for the "High Rate WPAN" technologies.

- **Task Group 4 (IEEE 802.15.4)**   This task group developed standards for the "Low Rate WPAN" technologies. The standard is well suited for wireless applications that require low data rates and minimal power consumption. The IEEE 802.15 standard is at the heart of various commercial implementations, such as the popular ZigBee. The ZigBee Alliance is a group of companies that maintain and publish the ZigBee specifications. The functions of the ZigBee Alliance are similar to those of the Wi-Fi Alliance in the WLAN world.

Summary

Even though it wasn't mentioned explicitly, standards help drive the multi-billion-dollar wireless industry. You should now understand the benefits and importance of standards, and you should appreciate the relationship between the regulatory organizations, technical organizations, and standards formulating bodies. You saw firsthand the results of the hard work put in by the technical organizations discussed in Chapter 1.

You also learned about some legacy, current, and future standards that directly affect the end products and quality of the products that we as wireless network administrators need to manage.

PART II | Hardware

CHAPTER 4 | Wireless Hardware: Client Side

Key Skills and Concepts

- Identify what constitutes client-side hardware.
- Understand the role and importance of the chipset in wireless hardware.
- Learn about some popular chipset manufacturers in the wireless industry.
- Learn about the different interface types for wireless client hardware.

C oncepts, theories, and standards are all important, but what good are they if we can't actually put them to work? In this part of this book, we'll look at some of the hardware that implements the concepts, theories, and standards that were covered in the chapters in Part I. This chapter and the next look at three broad categories of hardware: chipsets (which are found in client-side and infrastructure-side hardware), traditional client-side hardware, and infrastructure-side hardware.

Strictly speaking, this chapter discusses client-side components and their features. Our so-called "client-side hardware" includes things you might find in complete and finished products, such as wireless client stations, which are combined with other pieces to form a whole. In this book, a "client-side device" refers to any hardware that participates in a wireless network and meets any two of the following descriptions:

- Client-side devices are generally self-sufficient and are useful by themselves. This means that they can serve purposes other than wireless networking–related functions.
- The device is naturally selfish—it does not normally provide any type of service to other devices. It is concerned only with itself.
- The device might easily fit into a tote bag or a shirt pocket.

You will also see mentioned some hardware that is not strictly client-side or strictly infrastructure-side.

We begin this chapter by delving into the world of chipsets.

Chipset

In the information technology (IT) world, the chipset is a somewhat abstract concept—it is always there, but it can be difficult to pinpoint or define exactly. Even though we refer to a chipset as though it were one single hardware component, it is in fact a group of specialized integrated circuits (ICs) that are designed to work together in symphony to perform specific tasks. To confuse issues even more, a chipset can have different meanings, depending on the industry that uses the term—be they graphics, audio, or network communications industries.

In this book, chipsets used in the wireless network communications industries are the focus. These chipsets are used primarily for analyzing, interpreting, and manipulating radio signals—also known as signal processing. Specifically, we are interested in chipsets that implement the IEEE 802.11 and other wireless standards in one form or another.

Components of a Wireless Chipset

A chipset comprises several discrete components that work in unison to give the illusion of singleness. But that mythical illusion will be dispelled here.

Most chipsets used in wireless communication systems comprise separate integrated circuits that are responsible for handling Media Access Control and Baseband Processor (MAC/BBP) functions and the Pure Radio functions of a wireless communications system. On the whim of the chipset manufacturer, other miscellaneous functions may also be embedded into the circuits, such as hardware-based encryption.

Media Access Controller and Baseband Processor

The MAC hardware component controls and negotiates access to the transmission medium for the radio waves. The transmission medium in this case is the air or free space. "Baseband" is a term used in electronics to refer to the original and unadulterated group of signals (bits) originating from a data source before any kind of modulation takes place on the signal. The BBP hardware takes care of converting these baseband signals from digital to analog form (known as *modulation*) and then back again (known as *demodulation*) at the other end of the communication channel. Basically, the BBP is where modulation and demodulation take place.

The MAC/BBP component of a wireless communication chipset jointly performs the following functions:

- Interfacing with the host system via the applicable bus or interface type—such as via Peripheral Component Interconnect (PCI), CardBus, Mini PCI, or Universal Serial Bus (USB)

- Handles the analog-to-digital conversion (ADC)

- Handles the digital-to-analog conversions (DAC)

- Handles power management functions for the chipset

- Processes media access techniques, such as Carrier Sense Multiple Access with Collision Avoidance (CSMA/CA)

- Implements the modulation technique and the physical layer (PHY) properties of the IEEE 802.11 or other standard; for example, direct-sequence spread spectrum (DSSS, used in IEEE 802.11b), orthogonal frequency division multiplexing (OFDM, used in IEEE 802.11a/b/g), quadrature phase shift keying (QPSK), and differential phase shift keying (DBPSK)

Radio Frequency Transceiver

The integrated circuits in the radio frequency (RF) transceiver component of a wireless chipset handle the following functions:

- Reception of RF signals
- Transmission of RF signals
- Power amplification and low-noise amplification
- When applicable, provides an interface to the antenna

Figure 4-1 depicts the position of a wireless chipset in a wireless communication system.

Chipset Makers

The chipset industry is dominated by a few manufacturers (a vertical market), and every device manufacturer that requires the functionality provided by a chipset obtains it from one of these manufacturers. This is why so many different hardware brands from different manufacturers carry the same chipset, and it's why some wireless hardware drivers can be used across different hardware brands.

These device manufacturers add on a few extra customizations, stuff it all in nice cases, slap on their company logos, shrink wrap the products, include their markup, generate a lot of marketing material and thick manuals for the products, and then put them on the store shelves (horizontal market).

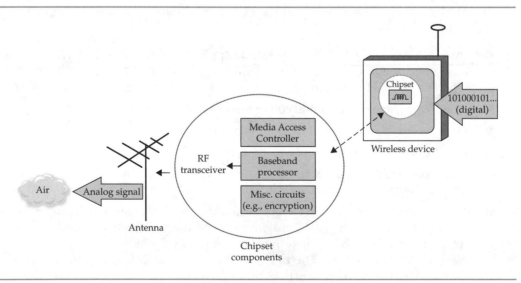

Figure 4-1. Chipset in a wireless communication system

Chipset Manufacturer	Chipset Code	Wireless Equipment Manufacturer
Atheros Communications	AR9002, AR9001, AR5008, AR5007, AR5006, AR5005, AR5002	Netgear, D-Link, and TRENDnet
Broadcom	Bcm43xx, BCM47xx	Apple, Belkin, Dell
Cisco Systems	Aironet	Cisco
Texas Instruments	acx111, acx100, acx100_usb	D-Link, US Robotics, Airlink, Netgear, Linksys
Alcatel-Lucent	Orinoco, Hermes	3Com, Apple, D-Link, Intel, Microsoft, Toshiba, Farallon, Avaya
Intel	Intel PRO/Wireless (IPW) 2200BG/2915ABG/ 3945ABG/4965AGN	Dell, Intel
Realtek Semiconductor	RTL818x, RTL8187B	Netgear, Belkin, D-Link, Linksys, Zonet
Ralink Technology	RT2500, RT2501, RT2600, RT2501USB, RT2800, rt2x00	Gigabyte Technology, Linksys, D-Link, Belkin, Nintendo

Table 4-1.   Sample Chipset Manufacturers

Some chipset manufacturers and companies that use the chipsets in their products are shown in Table 4-1. Note that the information in the table was current at the time of this writing.

NOTE   Some chipset implementations work in conjunction with firmware, and the functionality of the chipset is closely tied to the firmware. The firmware contains the main program to be run by a wireless card's chipset. This will be discussed in more detail later in the book.

Client-Side Wireless Hardware

Any piece of client-side wireless hardware always has at least one chipset embedded in it to perform various functions. Almost anything conceivable can be turned into a wireless client these days—from the humble toaster, to big and powerful computers. The toaster example may seem a bit far-fetched, but, truth is, as long as we can find a way to slap a radio on it, we can turn it into a node in a wireless network. It's all about the interface and how we intend to communicate with or over that interface.

A client-side device needs to participate in a wireless network. In wireless local area network (WLAN) lingo, such a device is often referred to as a WLAN *station* (STA).

Some client-side WLAN hardware may appear to be built into the host system, but you should understand that even when this is the case, the devices will still need to communicate with other hardware, using one or a combination of interface technologies. Now let's look at some common wireless client hardware interfaces, their incarnations, their form factors, and their bus interconnect types. These devices either provide an interface to the wireless chipset on the host system or have the wireless chipset embedded in them already.

PCI

Peripheral Component Interconnect (PCI) is a communication bus architecture over which PCI-based devices can communicate. Specifications that govern PCI are developed and defined by the PCI-SIG (PCI Special Interest Group), an international consortium.

Devices using the PCI architecture communicate by sending data signals simultaneously over parallel channels. Wireless hardware implementations using this interface often come in the form of wireless PCI expansion cards. They offer a cheap and easy way to allow most standard consumer PCs or servers to participate in a wireless network. Almost all modern computers still support this interface standard.

Wireless cards that use this interface often offer the added benefit of being able to use detachable antennas. This capability provides a cheap and easy way of boosting the reception capabilities of a WLAN device. Figure 4-2 shows a sample PCI wireless card.

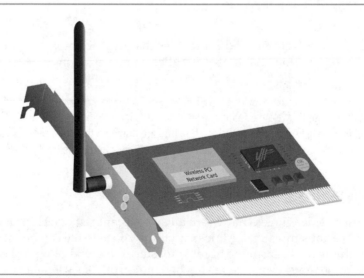

Figure 4-2.   PCI wireless card

Mini PCI

Mini PCI is a PC expansion interface standard. It is small form factor implementation of the PCI standard and is found primarily in laptops and other portable computers. Devices that use this interface communicate with the host over the PCI bus.

The Mini PCI standard is based on a subset of the same signal protocol and electrical characteristics as the main PCI specification. Hardware implementations of this standard are therefore functionally identical to full-sized PCI cards that are used in desktop systems.

Numerous hardware applications built around the Mini PCI standard are used for providing Bluetooth, WLAN, Ethernet, and modem support on host devices. Figure 4-3 shows a sample Mini PCI card.

Older laptops usually sport wireless cards that use this interface technology. Mini PCI cards have been superseded by PCI Express Mini cards, discussed next.

PCI Express Mini

PCI Express Mini (also known as Mini PCIe) is the PC expansion interface standard and an incarnation of the PCI Express standard. Specifically, it is a small form factor implementation measuring 30×56 mm, making them smaller than Mini PCI cards. Devices that use this interface communicate with the host either via PCI Express bus or via USB 2.0. The communication is via point-to-point serial links.

Wireless cards found in newer laptops are usually of the Mini PCIe type and connect via a Mini PCIe expansion interface on the host. Figure 4-4 shows a sample Mini PCIe card.

Figure 4-3.   Mini PCI card

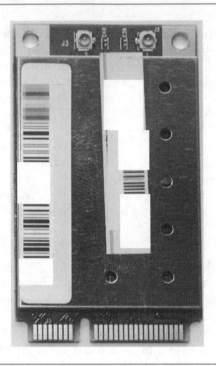

Figure 4-4.   Mini PCIe card

PC Card

The PC Card standard was developed by a group of companies originally known as the Personal Computer Memory Card International Association (PCMCIA). The group initially envisaged its standard to be an alternative to an existing standard developed by another group called the Japan Electronic Industry Development Association (JEIDA). The two groups later had a change of heart and decided to streamline the two competing standards.

Peripherals using this standard are designed especially with laptops in mind and also to fit into small spaces—Ethernet cards, WLAN network cards, modems, and so on.

The PC Card standard has gone through several (confusing) hardware generations based on several (more confusing) revisions. All generations of the hardware use a 68-pin connector and are all the same length (85.6 mm) and width (54 mm) but have varying thicknesses. Some PC Card standards and their associated form factors are listed in Table 4-2.

Standard	Supported Form Factor(s)	Features
PCMCIA version 1	Type I, Type II	Type I: ■ Implemented in older hardware ■ Supports only a 16-bit interface ■ Uses the hosts' ISA parallel bus ■ Designed mostly for memory card type applications and not input/output type applications, such as networking ■ Supports throughputs of up to 20 MBps ■ Cards are 3.3mm thick
PCMCIA version 2.0 JEIDA version 4.1	Type II	■ Merges competing PCMCIA and JEIDA standards ■ Support for input/output type applications ■ Support for dual voltage cards: 3.3 and 5 volts ■ Often must be used with dongles because the cards are usually no more than 5.5 mm thick ■ Supports throughputs of up to 7.84 MBps
PCMCIA version 2.01	Type III	■ The PCMCIA body started adopting the "PC Card" moniker ■ Come in 16- or 32-bit interfaces ■ Physically almost double the thickness (10.5 mm) of preceding Type II cards
PC Card version 5.0	CardBus	■ Adds support for the 32-bit CardBus interface ■ Operates at 3.3 volts ■ Supports throughputs of up to 132 MBps ■ Cards are easily identifiable by metal strip on top of their sockets ■ Slots are backward-compatible with preceding slot/interface types (Type I, II, III)

Table 4-2.   PC Card Standards

ExpressCard

ExpressCard is the next generation hardware standard developed to replace devices based on the PC Card technology. ExpressCard is based on a standard developed by the PCMCIA standards body.

ExpressCard hardware can interface with the host system via a serial interface, either over a USB 2.0 interface or a faster PCI express bus. Devices based on this technology support higher data bandwidth compared with the PC Card technology. Throughputs of up to 2.5 Gbit/s (gigabits per second) are supported in this standard when the PCI Express bus is used, and throughputs of up to 480 Mbit/s are supported when using the slower USB 2.0 interface. Devices using the interface sport a 26-pin connector.

ExpressCard offers several advantages over CardBus:

- Smaller form factor
- Doubles the data rates supported in CardBus
- Less power drawn from the host for more efficient power use
- Costs less to make and implement, due to elimination of an intermediary controller required in CardBus

A good comparison of newer ExpressCard and older CardBus technologies is shown in Figure 4-5.

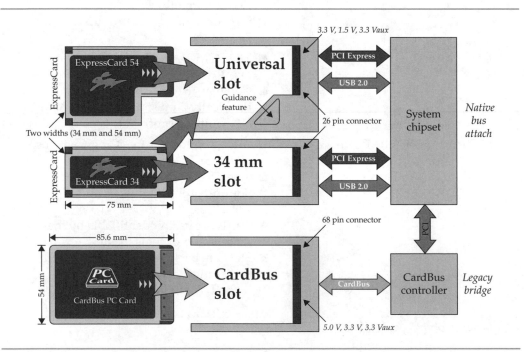

Figure 4-5.   ExpressCard technology vs. CardBus

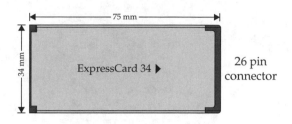

Figure 4-6.   ExpressCard 34

Peripheral cards that are built for the ExpressCard interface come in two form factors. The first form factor is rectangular in shape and is referred to as ExpressCard 34. This form measures 75×34mm (Figure 4-6).

The second form factor is L-shaped and is referred to as ExpressCard 54. It measures 75 mm at its longest length and 54mm at the widest width (Figure 4-7).

Most newer laptop models support ExpressCard interfaces. This standard is used in making PC peripherals for wireless WAN, IEEE 802.11 WLAN, and other applications that can benefit from the high data rates supported.

USB

Wireless USB devices are a popular method for adding WLAN capabilities to a host device. They are often in the form of a dongle (an electronic device that is usually attached to a computer externally) that can simply plug into any available USB port on the host system. Wireless cards designed to be plugged into the USB port of the host device often do not have any interface in which to plug an external antenna, as the antenna is often built into the device itself. A typical wireless USB adapter is shown in Figure 4-8.

You'll find plenty of wireless USB devices that implement virtually any of the individual IEEE 802.11 family of standards. Read the supported standards and platforms on the packaging box, and you should be good to go.

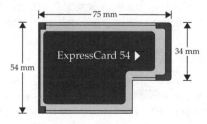

Figure 4-7.   ExpressCard 54

Figure 4-8.   Typical wireless USB adapter

WLAN devices that connect to a host via an external USB port are especially convenient for several reasons:

- Installing the hardware is easy and does not usually involve cracking the computer case open.
- They are not dependent on having an empty PCI, PCIe, or PCMCIA slot available.
- An abundance of USB ports are available on most host systems.
- The host system does not need to have any preexisting built-in wireless chipsets to become a wireless STA.
- They make it easy to enable most old or newer computer hosts to participate in a WLAN network.

 NOTE   In addition to their WLAN hardware applications, wireless cards using the USB interface are popular in devices for connecting to high-speed mobile broadband networks.

CompactFlash Cards and SDIO Cards

Portable devices such as personal digital organizers (PDAs), digital cameras, portable media players, mobile phones, and GPS receivers have now become an ubiquitous part of our lives and by extension part of the WLAN administrator's nightmare. You can add wireless capabilities to a lot of these devices cheaply and with little effort.

Portable device manufactures have lowered the costs of their devices by making Wi-Fi capabilities in their devices optional for the end user. The user can choose whether or not

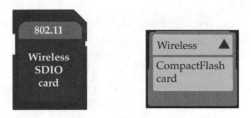

Figure 4-9. SDIO and CF cards

they require wireless functionality on the device. If the user requires this functionality, his or her choices to achieve this are limited only by the interface types and technology supported by the device vendor. In most cases, choices range from using the CompactFlash (CF) interface or a Secure Digital (SD) card interface on the host device.

In the wireless LAN (WLAN) world, CF cards are used for adding wireless capabilities to portable handheld devices (such as PDAs) that don't have this capability. CF cards are especially useful in this role because they require and draw very little power from their hosts, where battery conservation is already an issue.

Secure Digital Input Out (SDIO) is a memory card format, an extension of which can be used for various applications that include Bluetooth, Wi-Fi, and radio-frequency identification (RFID), to mention a few. SDIO cards connect to the host via the host's SD card. The SD card slot on support devices is normally used for extending the storage capabilities of the host, but certain devices are capable of using their SD card interface for communicating with SDIO cards. Figure 4-9 shows SDIO and CF cards.

Miscellaneous Converters

Popular in this category are Ethernet-to-wireless converters and serial-to-wireless converters that are used for providing wireless capabilities to dumb legacy devices, such as serial printers that have no wireless capabilities and no easy way to provide such capabilities using a software (driver) and hardware combination. These interface converters are usually stand-alone devices and do not often offer the wireless network administrator any graphical user interface (GUI) with which to configure them. Configuration of these devices is usually done via a serial port on the devices.

Other Considerations for Choosing Wireless Hardware

When making decision about which wireless client devices to use for a WLAN, you should consider several factors. One of these considerations is the already discussed obvious choice of the form factor and interface of client device (obvious, for example, because it is physically impossible to fit a Wireless PCI card into a the USB slot of a laptop). Other considerations for selecting wireless client devices are discussed next. Note that these considerations are also applicable to wireless infrastructure hardware.

Transmit Power (Tx Power)

Communication, as we all know, is a two-way street. We have the talking lane and the listening lane (let's ignore the existence of one way streets for now). In the WLAN world, this is akin to the transmitting and receiving of data that has been encoded as radio signals.

Transmit (Tx) power is relevant on the talking lane of our two-way street. It is the amount of radio frequency power that a transmitter produces at its output. Transmit power values are often expressed in milliwatts, but they can also be expressed as watts or decibel milliwatts (dBm). The transmit power of WLAN equipment is generally less than 1000 milliwatts (1 watt), and values are often represented as positive numbers.

The transmit power has a direct relationship with the range of an RF signal. In other words, the higher the transmit power, the further the signal can travel to its intended (and unintended) destination. As discussed in Chapter 2, the property of a radio wave that affects its power is its amplitude.

Receiver Sensitivity (Rx)

This characteristic is relevant on the listening lane of our two-way street—the receiving end of the communication. The receiver sensitivity is a somewhat tricky measure, and few device manufactures bother to supply hard figures for this characteristic. It is a measure of the minimum threshold at which a WLAN client device can properly detect and interpret a radio signal.

The unit of measurement for the receive sensitivity of a wireless device is the dBm. The measurement values are often expressed as negative numbers, so a wireless device with a low receiver sensitivity value (such as –90 dBm) can receive and better interpret radio signals than another device with a higher receiver sensitivity value (such as –70 dBm). In other words, the lower the receiver sensitivity, the weaker the signal the wireless device can pick up—which is a good thing.

Summary

The chipset is integral in the grand scheme of almost all wireless communications hardware. Some big players in the chipset world use various hardware interfaces and form factors to create these chipsets in all shapes and forms.

Client-side wireless hardware comes in different types and with different interfaces. The interface used depends on the available interface on the host device and also on how we intend to communicate over that interface.

Amongst other things, a wireless network administrator must consider two important RF measures when selecting wireless client hardware: the transmit power and the receiver sensitivity.

The next chapter discusses some of the hardware used on the infrastructure side of wireless networks.

CHAPTER 5 | Wireless Hardware: Infrastructure Side

Key Skills and Concepts

- Understand what constitutes infrastructure-side wireless hardware.
- Understand the function of an antenna in wireless communications.
- Learn the various concepts associated with antennas.
- Identify different antenna types.
- Understand the functions of wireless access points and residential gateways.
- Understand the functions of wireless controllers and wireless bridges.
- Learn the basics of Power over Ethernet.

Chapter 4 discussed the wireless hardware found on devices that serve as clients on a wireless network. This chapter covers the hardware that makes up the infrastructure end of things. These hardware can individually and collectively connect end nodes (clients) on a wireless network. Some of the hardware and hardware interfaces used in wireless clients are also present in infrastructure devices. For example, the same types of chipsets used in client hardware are often used in the infrastructure devices as well.

Broadly speaking, what this book calls an "infrastructure-side device" is any hardware that participates in a wireless network that meets any two of the following criteria:

- Helps to connect other wireless devices together
- Provides a service to a wireless client
- Will not easily fit into a tote bag

As you will discover later, some devices fall into a "gray area" regarding these criteria—but read on and you'll understand why this is so.

Antenna (Aerial)

In Chapter 4, you read that antennas (also called aerials) are usually built into wireless radio cards used on the *client* side. But those types of antennas are small and nice-looking. In this chapter, when antennas are mentioned, think big. Think massive. Think antennas that need to be positioned 300 feet above sea level. Antennas that are 500 feet in diameter. Think satellite dishes. Seriously speaking, antennas come in all shapes and sizes. A manufacturer designs its antennas to suit specific applications. If the antenna is meant to fit into a small ExpressCard form factor, it has to be very small. On the other hand, if it is meant to link two remote buildings together, it will be very large (and will most likely not fit into a tote bag).

An antenna is the first port of entry and exit for the radio signals passing through a wireless communications system. It is the hardware from which the radio waves emanate and to which radio waves aggregate.

The electronics in any wireless hardware send electrical signals to the antenna, and the antenna converts this electrical current into electromagnetic waves, which then propagate through the air to the final destination. This destination is usually another antenna, which may be embedded in a wireless client hardware or some other wireless infrastructure device. After they arrive at their destination, the electromagnetic waves are converted back into electrical signals.

Antenna Concepts

Several common terms and concepts are important to know regarding antenna characteristics. These concepts and terms will help you understand discussions on antenna types.

Intentional Radiator

The intentional radiator (IR) is any hardware component of a wireless communications system that intentionally generates and emits radio frequency (RF) energy. By definition, the IR is distinct from the antenna. The IR consists of components such as the transmitter, amplifiers, cables, and other knobs and dials found in wireless devices. The IR ends where the antenna starts. Cordless phones, walkie-talkies, and wireless cards are all good examples of IRs (excluding the antenna portion, of course).

An IR is in contrast to an unintentional radiator, which does not deliberately generate RF energy but does generate it as a by-product. Examples of unintentional radiators are electrical transformers and dynamos.

Regulatory bodies within each country govern the maximum RF power levels that an IR can produce. One of the reasons this is necessary is for safety—that is, we don't want humans exposed to too much RF energy. This is because the exact long-term effects of RF energy on human health is unclear. In addition, RF-generating devices can sometimes interfere with the proper functioning of other electronic devices, like lifesaving equipment, such as pacemakers.

Equivalent Isotropically Radiated Power

Equivalent isotropically radiated power (EIRP) is a measure of the amount of actual power emitted from an antenna. The EIRP acronym sometimes represents effective isotropically radiated power. The *isotropy* in EIRP describes an ideal situation whereby the radio energy is assumed to be equally distributed in all directions. This type of ideal situation, of course, does not exist in the real world.

EIRP is a measurable quantity. It is the sum total of the power output from the IR and the passive gain (or loss) caused by the antenna. Regulatory bodies within each country also govern the values of the EIRP of wireless devices.

The following illustration shows the IR and EIRP components of a simple wireless station (STA).

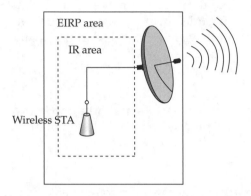

Antenna Gain and Loss

An antenna can cause either a gain or a loss in the RF signal it is transmitting or receiving. The type of gain or loss caused by an antenna is passive in nature, in the sense that the antenna does not actively cause gain or loss in RF signal strength by using any special electronics (such as amplifiers or attenuators). Instead, an antenna can cause signal gain because it helps to focus or concentrate the RF signal in a specific direction. And, conversely, an antenna can result in reduction in the effective signal strength by simply unfocusing it.

The following illustration depicts the before and after effects of introducing a high-gain antenna into the communication channel between wireless STAs—WAP 1 and WAP 2. It shows how an antenna can produce gain in the RF power by making the signal more focused.

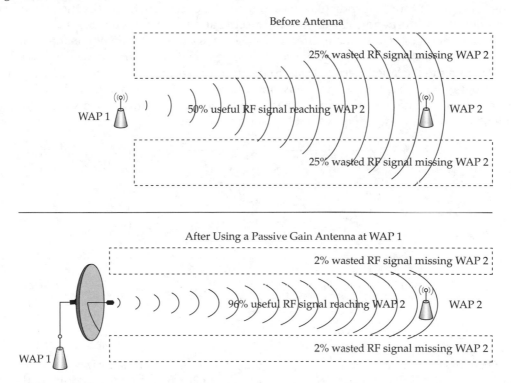

Beam Width

Beam width is a measure of the horizontal and vertical reach of the RF signals transmitted from an antenna. Beam width is measured in degrees. The vertical beam width is perpendicular to the ground, and the horizontal beam width is parallel to the ground.

The beam width specification of an antenna is important in choosing an antenna that will provide the coverage needed for a specific area. Practically speaking, this could mean that in some situations, an antenna with a narrow beam width is more desirable than an antenna with a wide beam width.

Polarization

An electromagnetic wave comprises two components: an electric component and a magnetic component.

The electric component is also called the *E-plane*. The E-plane is parallel to the conductor, or element, of the antenna. The magnetic component is called the H-plane. The components are perpendicular to each other. Electromagnetic waves travel in a direction that is perpendicular to both the electric and magnetic components.

Polarization of an antenna is affected by the physical orientation of the antenna.

When the antenna is oriented vertically, the polarization is vertical, which also means that the E-plane is *perpendicular* to the ground. This is the case, for example, with the types of antennas that are normally oriented vertically, such as in omnidirectional antenna (more on this later in this chapter). The following illustration shows a vertically oriented antenna with the equivalent E-plane and H-plane:

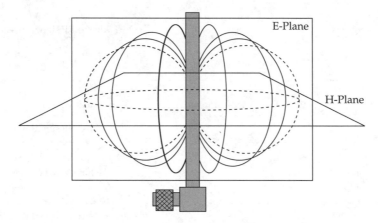

When the antenna is oriented horizontally, the polarization is horizontal, which means that the E-plane is *parallel* to the ground. This is the case, for example, with the types of antennas that are normally oriented horizontally, such as yagi antennas (more on these a bit later). The following illustration shows a horizontally oriented antenna with the equivalent E-plane and H-plane:

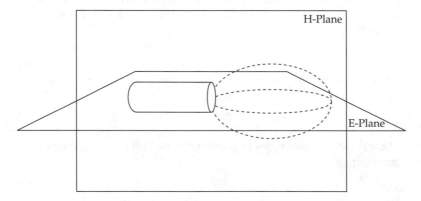

In a wireless communications system, the transmitting and receiving antennas should be polarized in the same direction to achieve best results.

Antenna Types

The shape and properties of an antenna are determined by the application and environment in which it will be used. Antenna types are categorized as omnidirectional, semidirectional, and directional.

Omnidirectional Antennas

Omni means *all*. Omnidirectional antennas transmit signals in all directions and receive signals from all directions. Omnidirectional antennas transmit and receive RF signals in 360 degrees. The horizontal beam width of the RF signal emanating from an omnidirectional antenna is 360 degrees, and the vertical beam width is usually between 7 and 80 degrees.

These are probably the simplest, cheapest, and most common types of antenna available. They are frequently found in commodity wireless hardware used on the client and infrastructure sides. A dipole antenna is an example of an omnidirectional antenna.

The wave pattern of the radio waves transmitted and received from omnidirectional antennas is said to be shaped like a *toroid*, or a ring donut, as shown in the following illustration:

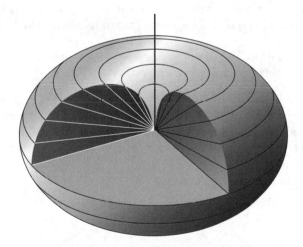

Omnidirectional antennas are best suited for use in environments such as the center of a room or space because of the 360-degree horizontal coverage they offer. The signal emanating from omnidirectional antennas propagates more horizontally than vertically.

Whenever coverage is needed in relatively equal amounts all around a horizontal space, an omnidirectional antenna should be considered. Just remember that they are not suited for environments in which vertical coverage is more desirable, such as the lower and upper floors of a building. Omnidirectional antennas are often used for indoor applications for which radio signals are transmitted from one device to multiple devices at the same time—called point-to-multipoint or one-to-many systems.

Numerous stand-alone omnidirectional antennas are designed to be attached and used with wireless client devices such as PCMCIA cards, PC Cards, and PCI Cards.

Semidirectional Antennas

Semidirectional antennas radiate and receive RF signals in a roughly 180-degree beam width. Several types of semidirectional antennas are available. When used indoors, semidirectional antennas can be placed on a wall facing the desired direction or mounted on the ceiling facing down.

In terms of application, outdoor semidirectional antennas are best suited for short-to-medium haul transmissions and receptions.

In terms of placement, these antennas are best placed in locations where signal reception is needed in one direction and not so much in the other, such as at the end of a long hallway with offices on either side.

The wave pattern of the radio waves transmitted and received from semidirectional antennas is shaped like a ring donut cut in half.

Planar Antennas   Planar antennas are a special subcategory of semidirectional antennas. They do not have the conventional antenna physical structure. Because of their internal design and the flexibility offered as a result, planar antennas can come in various shapes and sizes. They are usually deployed for indoor applications and are not often easy to spot because they can be packaged so that they blend in with the rest of their environment.

As diminutive and harmless as they appear, planar antennas are used for very serious applications, such as cellular, broadband, WLAN deployments.

Yagi   A yagi antenna is commonly used for outdoor applications. They also come in several shapes and sizes, but a distinctive physical feature is that they usually have some kind of protruding snout that emits the RF signals. As with most other types of antennas, a yagi antenna's signal gain comes from its ability to limit the horizontal and vertical beam width of the RF signal.

Figure 5-1 shows a sample yagi antenna.

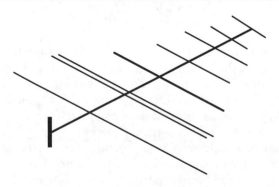

Figure 5-1.   Yagi antenna

Directional Antennas

Directional antennas focus their radio waves in a very specific single direction. The horizontal beam width of the RF signal emanating from a directional antenna is between 4 and 25 degrees, and the vertical beam width is usually between 4 and 21 degrees. These antennas generally provide high signal gain, and installation and handling should be undertaken with great care to avoid undue exposure to high-power RF radiation.

The wave pattern of the radio waves transmitted and received from directional antennas is shaped almost like a cone.

These antennas are best placed in locations with a clear line of sight to the receiving end of the communications system. Most often, the receiving end will also be another directional antenna. This makes these types of antennas suitable for point-to-point applications. Directional antennas are best suited for medium-long haul transmissions and receptions.

NOTE It might seem like these antennas do a perfect job of focusing their radio waves in the direction for which they are designed, but this isn't specifically true. All antennas propagate some RF energy in some unintended directions, even though an important antenna design objective is to limit this unintended transmission as much as possible.

Other Antenna Considerations

The placement and type of antenna used can be the defining factors of good signal transmission and reception for wireless communications. Even though most wireless devices ship with their own antennas, a wireless network administrator sometimes needs to swap out the original antenna and replace it with one that works better for a specific situation. In addition, numerous wireless device manufacturers' products allow the end user to choose the most suitable antenna for specific applications.

It may be necessary for the wireless network administrator to consider wireless antennas in the following situations:

- The default antenna is not adequate for transmitting the desired signals to the destination.

- The default antenna is unable to receive the desired signals.

- The antenna is causing the wireless device to violate local wireless regulatory rules.

- The wireless device does not ship with an antenna, but includes a connector for attaching an appropriate antenna.

Antenna Checklist

If the wireless network administrator determines that an antenna is needed for a specific application or scenario, it is important to select the best antenna for the job.

Several factors should be taken into consideration in antenna selection. These would include polarization, beam-width, antenna gain, antenna loss, as well as the following additional considerations:

Outdoor vs. Indoor Antenna   Antennas that are designed for mounting outdoors are different from indoor antennas. The differences are not necessarily in the electronics or functionality of the antenna but are often in the shape, protective enclosure, and other physical characteristics of the antennas that are designed for outdoor use. In general, you would want an outdoor antenna to be weatherproof so that the electronics are well-protected come rain, snow, or sunshine.

Aesthetics   When it comes to antennas, looks can matter. The enclosures of some antennas are designed so that they blend well with the interior or furniture in a space. Some antennas are designed to be as inconspicuous as possible.

Mounting Kits   This is an important consideration when selecting an antenna. If an antenna needs to be mounted on an external surface, you must carefully choose the parts used for the mount. Most antenna manufacturers supply mounting brackets and kits that work with their hardware. But the manufacturer's mounting kit might not be suitable for your application; in this case, you'll need some third-party solutions for the mounting pieces. For example, the mounting parts needed for mounting an antenna on a wall will be different from those needed for mounting an antenna on a pole or a ceiling.

Placement   The actual placement of the antenna is another important consideration. Indoor antennas need to be placed in a location that will receive minimum interference from other electrical devices; immovable objects, such as walls, ceilings, or signs; or even humans. Sometimes, unavoidable environmental factors can affect the radio communications for which the antenna is being used.

Outdoor antenna placement can also be critical. Local regulatory laws regarding outdoor structures may also apply, and these laws can influence the total height, size, and appearance of an outdoor antenna. Make sure that you are not breaking any local laws with the antenna.

Don't forget that the antenna also needs to be easily reached should it require service or repair. Place the antenna so that it is easily accessible.

Safety   Proper grounding procedures should be followed when installing outdoor antennas, which are exposed to lighting and other natural phenomena and are directly connected to indoor electrical devices. Indoor antennas should be placed where they pose no dangers to humans, including physical or overexposure dangers, or where they could interfere with radio signals used on lifesaving equipment, such as equipment used in hospitals.

Wireless Access Point

A wireless access point (WAP) provides a means for wireless nodes (also known as Wireless Stations) to communicate with other wireless nodes or to communicate with a wired network. A WAP is often simply referred to as an access point (AP). They are commonly used in enterprise-type environments where the functions of networking hardware must be properly segmented—that is, these devices do not typically perform all functions but are instead designed to perform specific functions very well.

Traditional WAPs are somewhat distinct from wireless residential gateways or wireless routers because they do not normally perform Open System Interconnection (OSI) Layer 3 type routing functions. They instead rely on other components on the network to perform such functions.

The line between the role of WAPs and wireless gateways is getting more and more blurred with time as wireless network equipment manufacturers are consolidating all these functions into single-purpose solutions. There are advantages and disadvantages to this. One disadvantage is that the single AP acts as a single point of failure in the network. This means that if any of the components or functions fail, the device could bring down the entire network. On the other hand, having major functions of the wireless network available in one multipurpose device can make the wireless network administrator's job a lot easier.

You can identify traditional WAP hardware by its lack of a distinct wide area network (WAN) interface. Instead, it usually has only local area network (LAN) interfaces with which it connects to the LAN segment.

In addition to connecting wireless clients to a wired network, WAPs can also feature a combination of the following: security functions, MAC layer filtering functions, protocol layer filtering, VLAN functionality, detachable antennas, field replaceable radio cards, and Power over Ethernet, to mention a few. Some of these functions are covered in more detail later in the book.

WAP Operational Nodes

Wireless APs use three operational modes: root mode, the most fundamental and natural mode in which a WAP operates; bridge mode, used for bridging two or more networks together; and repeater mode, which helps to extend the reach of a wireless network. Figure 5-2 shows the three operational modes of APs.

Root Mode

In this mode, also called infrastructure mode, a WAP is performing its basic function—connecting wireless clients to a wired network. Root mode is also used for intermessaging or management purposes when APs need to communicate among themselves (for example, when two or more APs are combined for STA roaming purposes).

Bridge Mode

WAPs operating in bridge mode can be used to connect two or more wired networks together. The bridged network normally ends up being on the same subnet and shares a common broadcast domain. When operating in this mode, wireless clients are usually unable to connect wirelessly to the AP because the wireless interface of the AP is usually dedicated to performing the bridging function.

APs operating in bridge mode are popular for creating point-to-point and point-to-multipoint links.

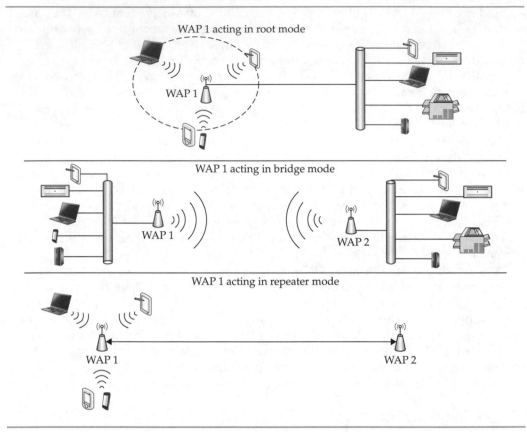

Figure 5-2.   All three operation modes of APs

Repeater Mode

In repeater mode, the WAP extends the reach of the wireless network by repeating the signals of a remote WAP. The repeater AP can be connected via a hardwire to the remote root mode AP. Wireless clients can then connect wirelessly to the WAP operating in repeater mode. This is one of the key differences between the AP repeater mode and bridge mode. The wireless client node can connect to a WAP repeater, and wired clients can connect to the WAP bridge. Note that the repeater functionality in this case should not be confused with the idea of traditional repeater, which generally involves signal amplification and regeneration.

Implementation of repeater mode in APs is vendor-specific and does not rely on any IEEE 802.11 standards. Because the implementation of the repeater mode is vendor-specific, it's a good idea to make sure that all wireless hardware used comes from the same equipment manufacturer.

When to Use a WAP

How does the wireless network administrator know when to use a WAP? Some guidelines can help answer this question:

■ The wireless network will span more than three contiguous spaces.

■ A solid wired network infrastructure is already in place, and the extra services of multipurpose or multifunction APs are not needed because the existing wired equivalents will be used.

■ Wireless switches or controllers will be in use. Other alternatives will not meet the technical needs or requirements that the WAPs are designed to provide.

Wireless Switches and Controllers

Wireless switches and controllers (WCs) are a class of infrastructure devices for which there is no ambiguity whatsoever in where they belong or where they are used in wireless deployments.

WCs are strictly an enterprise class of wireless hardware. They are designed to scale to meet the needs of different sized networks and serve as a sort of central management point for other STAs in a wireless network. So instead of your having to configure or manage WAPs, wireless routers, bridges, and other STAs in a wireless network manually and individually, you can do it all centrally from the WC. This improves network management efficiency.

In addition to STA configuration, the following wireless network administration tasks can be performed via a WC:

■ Software or firmware management for the STAs (upgrades, downgrades)

■ The individual configuration or settings for client STAs can be stored or saved on the WC, making it a configuration repository for client STAs

■ MAC layer filtering, protocol layer filtering, wireless intrusion detection capabilities

■ Quality of service management

■ Power over Ethernet (PoE) support for remote devices

■ Authentication and authorization management for connected clients

■ Dynamic Host Configuration Protocol (DHCP), routing services

■ AP load balancing management

Figure 5-3 shows a sample wireless controller and various examples of how other wireless STAs and wired nodes might connect to it in a network.

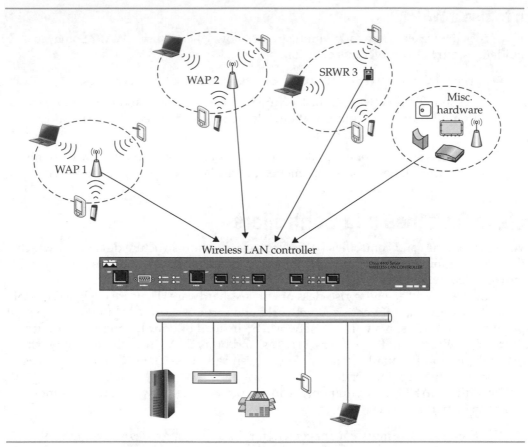

Figure 5-3.   How wireless STAs and wired nodes might connect to a wireless controller

Wireless Controller Considerations

Wireless controllers are often used in very large wireless networks. A good wireless network administrator will often intuitively know when one of these devices is required. Nevertheless, you can use the following to aid in deciding whether or not you need to deploy a wireless controller:

- Sales people from your favorite original equipment manufacturers (OEMs) keep calling to let you know that you need a controller to go with the 200 APs you purchased.

- You are managing a wireless network whose APs are distributed throughout different geographical locations.

After you have determined that a wireless controller is needed on your network, the following checklist can help you decide which type of controller to purchase:

■ Choose a controller that implements and supports common standards and protocols. Devices that implement too many proprietary technologies should be generally avoided.

■ If proprietary technologies cannot be avoided, try to stick with hardware components from the same manufacturers. This can help reduce compatibility issues that can arise between equipment from different vendors.

■ Modular controllers should be used whenever possible. This makes it easier to upgrade and provide support for newer technologies as they are developed in the future. In other words, it helps to future-proof your hardware investment.

Wireless Routers: SOHO and Residential Wireless Gateways

Small office home office (SOHO) and residential wireless gateways or routers are the common folk's answer to the enterprise-grade WAP discussed earlier. SOHO/residential wireless routers (SRWRs) are often mistakenly referred to as APs by lay people, and although they do provide most of the same services that the WAPs offer, they often provide stripped-down equivalents of the services with smaller footprints.

 NOTE SRWRs aren't the only kinds of wireless routers available. Wireless OEMs also manufacture a wide variety of enterprise-grade pure wireless routers. But SRWRs are the most prevalent class of wireless routers available today.

Residential gateways normally possess much less processing power (CPU, memory, and so on) than their enterprise-grade counterparts. But they have enough power to serve their primary function: to provide wireless and wired nodes in a network access to another network—the other network usually being a larger network, often a WAN (such as the Internet). Some routing logic is required in the device before it can link two or more networks on different subnets. Physically, wireless routers are distinct from APs because of the presence of a so-called WAN interface.

Wireless routers today provide a plethora of services and can have numerous features, such as the following:

■ Support for different security mechanisms, such as MAC layer filtering and protocol layer filtering via built-in stateful packet inspection (SPI)

■ Captive portals

■ Network attached storage hubs

■ Virtual local area network (VLAN)

Open Source Alternatives

Several mature and thriving open source communities abound that specialize in extending the stock features of several SRWRs. Some of the features provided by these (often free) alternatives are pretty mind-blowing and can make you wonder whether OEMs are not deliberately crippling their hardware by not shipping or supporting the features in their own devices out of the box.

You can often improve the feature set of these wireless hardware via software and firmware upgrades. Some small hardware modifications might be required, but this is rare. You can find more information about these alternatives at these URLs:

- www.openwrt.org
- www.freewrt.org
- www.dd-wrt.com
- www.polarcloud.com/tomato

Although most of these kinds of web sites include standard disclaimers about using their products on your hardware, you can be assured of the high quality and usefulness of these alternatives.

- Switching functions
- Routing functions
- Quality of service (QoS) functions
- Domain Name System (DNS) server
- DHCP server.

It is not uncommon to find SRWRs that can operate in the different WAP modes—root mode, bridge mode, or repeater mode. In fact, you would be hard pressed to find a wireless router today that does not support or provide almost all the features listed here.

Wireless Bridge

A wireless bridge is a pretty much a stripped-down WAP operating strictly in bridge mode. Wireless bridges are dedicated to performing this bridging task and often lack most of the other bells and whistles that full-functioning WAPs or SRWRs posses. They are used purely for linking two separate wired network segments together wirelessly. Wireless bridges are not normally designed to be connected to directly by any wireless client devices.

These bridges are typically deployed in one of two fashions. First, they can be used to link two remote sites together wirelessly, which is called a *point-to-point* (one-one) link. In this deployment, a clear line of sight (LOS) is normally required between the two end points. Second, they can be used to link three or more sites together, which is called a *point-to-multipoint* (one-many) link.

In either of these deployment types, one of the bridges acts as the master or root of the link and the other(s) act as the slave(s) or non-root. The master (root) is the central and most knowledgeable node in the link. All the other nodes communicate only through and via the root bridge and not directly with one another.

 NOTE   Another wireless bridge deployment type is used for setting up redundant links or load balancing traffic between two remote locations wirelessly. To avoid the confusion (loops) that can occur when more than one path exists for packets while traversing the network, a Spanning Tree Protocol (STP) mechanism is required. Four or more wireless bridge devices can be combined to create this deployment. If a failure occurs in one of the links, redundancy is provided because traffic between the sites can failover to the other good bridge link. Load balancing or load sharing can be used to specify the type or nature of the traffic that can be transferred over each individual bridge link. For example, the WLAN administrator can specify that all voice traffic be transferred over one link, while all non-voice traffic be transferred over the other link.

Power over Ethernet

Power over Ethernet (PoE) is not exactly a type of infrastructure hardware. It is a technology that is sometimes found implemented in wireless devices used on the infrastructure side. PoE is also not specifically a wireless-centered technology, and as such it is not governed by the IEEE 802.11 wireless standards. Instead, it's described in a clause in the IEEE 802.3 standard as a method to transfer electrical power, as well as data, to remote devices over the common twisted-pair cable in an Ethernet network.

Hardware components implementing PoE can be grouped into two types: power sourcing equipment (PSE) and powered devices (PD).

PoE is especially useful to wireless network administrators because it is often necessary to position wireless networking equipment (such as APs, wireless routers, and switches) in awkward locations. One such location might be an unused part of a building that has no existing electrical wiring or outlet with which to power the hardware. The idea is that if we can somehow manage to get an Ethernet drop to the hardware, we can give it power as long as it supports PoE. It might be easier to create or extend a Ethernet drop instead of creating new electrical wirings and outlets. Using PoE, we can easily place a wireless STA far up in the ceiling of a warehouse or in the depths of the Amazonian jungle.

Additional Considerations

When you're making a decision about which hardware to use in a wireless network infrastructure, you need to take several other factors into consideration: specifically, cost and interoperability.

Cost

Cost is usually a no brainer because, for most of us, financial resources are limited. Wireless network administrators and managers are required to work within a given budget range, and a delicate balancing act is involved in keeping wireless network equipment capital costs down, while continuing to maintain equipment and support service agreements, staff training, and so on.

Need is probably the most important single driving factor that can affect the decision of whether to make the equipment purchase or not when budgets are tight. If you need it, you need it. And if you can do without it, then you don't need it. Small, medium, and large businesses these days are trying to run as thinly and efficiently as they can to stay afloat and profitable.

Interoperability

Interoperability refers to the way some equipment (wireless network hardware in our case) works with other equipment. It deals with compatibility issues as well, which may not always be clear-cut. Incompatibilities can stem from the way vendors (mis)implement the IEEE 802.11 family of standards in their products. It can also result from an OEM's desire to corner a certain market segment by deliberately making its hardware and the protocols used proprietary.

Interoperability issues can also arise for completely innocent reasons, such as when you're trying to work with legacy equipment or protocols. It is not often that the wireless administrator or network designer has the opportunity to design or work with a brand new network from scratch. We often have to deal with and accommodate existing or old hardware and clients.

The only way to mitigate most interoperability issues is by careful planning and *extensive* testing before deploying infrastructure-type hardware. A lot of real-world snafus can be discovered during testing, and this will ultimately aid you in making the right decision.

Summary

Infrastructure-side wireless hardware includes a variety of hardware that might be difficult to categorize as client- or infrastructure-side hardware. When this hardware is used on the client side (such as the antenna), it can be quite nondescript (and taken for granted), but you know that it plays an important role nonetheless in any wireless communications system.

Some hardware is clearly used only on the infrastructure side of a wireless network. Various categories of hardware can aid connectivity and provide one service or the other to other wireless STAs or wired nodes in a network. You need to consider several important factors when dealing with infrastructure hardware, which were covered in this chapter. You also learned about PoE, a technology that allows you to power *any* supported wireless networking equipment almost anywhere.

PART III | WLAN, WWAN, WMAN, and WPAN

CHAPTER 6 | WLAN

Key Skills and Concepts

- Define common wireless local area network (WLAN) terminologies.
- Examine the key components of a WLAN.
- Understand the IEEE 802.11 standard governing WLANs.
- Understand the physical and Media Access Control (MAC) components of IEEE 802.11.

A local area network (LAN) spans a relatively small physical area—*relatively small*, because the area can be the size of a closet, a room, a home, an entire office, or a group of buildings. Compared to a wide area network (WAN), such as the Internet, and a metropolitan area network (MAN), a LAN seems pretty small. Several LANs can be used to make up the much larger networks of WANs and MANs.

A WLAN is a type of LAN technology. In fact, the IEEE 802.11 standard that governs the basic operations and functions of WLANs says that a WLAN is required to appear to higher protocol layers as a regular wired IEEE 802 LAN.

WLAN Components and Concepts

WLANs comprise several components, some purely logical and others physical. The following sections introduce some of the components that make up a WLAN and define some of the lingo used in describing the entities in a WLAN.

Wireless Medium

The *wireless medium* (WM) is used for the actual transfer of information between the entities of a WLAN—that is, air and space.

Various types of data can be encoded into a type of electromagnetic disturbance called radio waves. These radio waves are then transmitted over the air (our medium) to their destination, where they are decoded back into useful data.

Wireless Station

Any device that implements the IEEE 802.11 standard is called a *wireless station* (STA).

A STA is therefore a single physical entity that can perform the 802.11 song and dance. Wireless STAs are not very useful by themselves; they need other wireless or wired devices to send information to and receive information from in order to make them useful.

Figure 6-1 shows some sample wireless STAs.

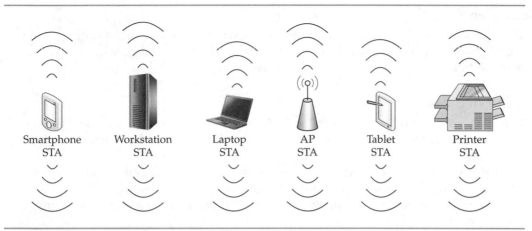

Figure 6-1.   Sample STAs

A STA can operate in one or more of the following modes:

- **Infrastructure mode**   Also referred to as client mode STA, this mode requires the use of an infrastructure device, such as an access point (AP), for communicating with other STAs, as well as with the wired LAN. Devices operating in this mode can be considered wireless clients, which implies that the wireless STA acts as a client in a WLAN. Other literature may refer to client mode STAs as *mobile units* (MUs). Despite the name, MUs can, of course, be either mobile or in a fixed location. Infrastructure mode is probably the most typical and common mode of operation for the vast majority of wireless devices.

- **Ad hoc mode**   Ad hoc STAs form autonomous networks that do not require an AP to communicate with other STAs. They can be either mobile or in a fixed location.

- **Access control mode**   As implied by the name, an access control STA is used for controlling access between STAs or controlling STA access to the wired LAN. A wireless access point (WAP) is a good example of a STA that operates in this mode.

Distribution System

The *distribution system* (DS) is an important part of any network and serves as a "glue" for interconnecting similar or dissimilar networks together.

The fabric of the DS can be based on any of the common wired technologies such as Ethernet. It can also be entirely wireless using regular wireless technologies such as those based on IEEE 802.11.

In wireless networks, the DS can be used for linking the wireless STAs to the wired resources. In fact, the most common use of the DS is for bridging a wireless network to the wired network.

Access Point

The *access point* (AP) is a type of STA. Specifically, it helps link wireless stations to the wired stations or resources, or it may simply be used for connecting wireless STAs to one another.

Basic Service Set

The *basic service set* (BSS) is a logical entity in a WLAN. The BSS can also be viewed as a type of WLAN topology. BSSs come in two "official" types: *independent* BSS and *infrastructure* BSS.

A third, but "unofficial," topology type exists, called a *wireless distribution system* (WDS). It's unofficial because it is not officially sanctioned or described in the 802.11 specification, and neither is it sanctioned by the Wi-Fi Alliance special interest group. But it is widely used in WLAN setups.

Infrastructure BSS

Infrastructure BSS, frequently referred to simply as a BSS, makes use of an AP. Most WLANs operate in this mode. The STAs that participate in the infrastructure BSS do not communicate directly with each other; all communications go through the AP.

The IEEE 802.11 standard defines the BSS as "a set of stations (STAs) that have successfully synchronized amongst themselves using the appropriate functions." These functions are called *primitives* in the 802.11 technical jargon.

Infrastructure mode STAs operate in infrastructure BSS. The following illustration shows a sample infrastructure BSS:

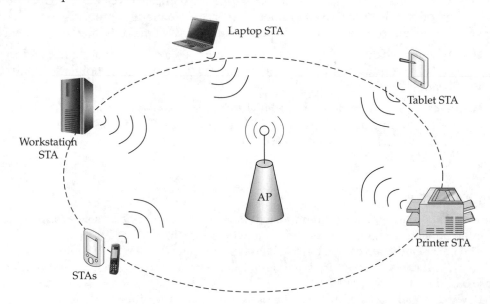

Independent BSS

The *Independent BSS* (IBSS) does not make use of an AP; the STAs communicate directly with one another in a peer-to-peer fashion. A minimum of two STAs are required to form an IBSS.

Ad hoc mode STAs operate in IBSS mode. The following illustration shows an IBSS:

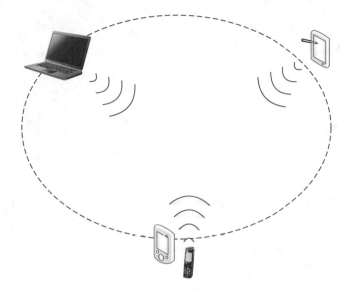

Wireless Distribution System

Wireless distribution systems (WDS) are used to describe WLAN topologies in which APs are connected together. Strictly speaking, it means that the infrastructure devices (or APs) are linked together using a wireless medium (air or space) as the distribution system. A WDS is used for creating a wireless backbone link between the APs in a WLAN. This is in contrast to the traditional method of linking the APs in a wireless network via a wired distribution system.

The WDS often requires that all participants employ and share various characteristics, such as a common radio frequency (RF) channel and a common security mechanism.

The next illustration shows a sample WDS.

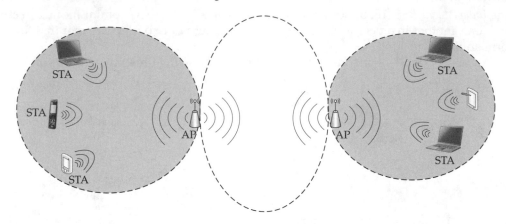

Service Set ID

The service set ID (SSID) is used to identify an extended service set (ESS) or IBSS. Specifically, it is a human-friendly means of identifying the ESS or IBSS. The SSID is the name that users often see when they are presented with a list of available wireless networks detected by their wireless device.

The SSID can be 0 to 32 bytes long.

Basic Service Set ID

It only makes sense that if we have a human-friendly way of identifying wireless networks, we should also have a machine-friendly way of doing the same thing. This is the job of the BSSID, which is used for identifying each BSS. It is 48 bits long and is very similar to the MAC address used on Ethernet-based networks.

The exact value of the BSSID depends on the service set in use (Infrastructure or Independent).

In the Infrastructure BSS, the BSSID is easily determined, because it is the MAC address currently in use by the wireless STA that is acting as the AP. Recall that the Infrastructure BSS always has an AP present. This MAC address is a universally administrated type, and it is always bound to be unique. A sample BSSID for an infrastructure BSS is 00:ab:34:56:78:9a.

In the ad hoc, or independent mode, BSS, the BSSID is a locally administered (non-universally unique) type of MAC address. A locally administered MAC address is one in which the value of universal or local (U/L) field bit is set to one (1). Specifically this means that the second least significant bit of the most significant byte of the MAC address has a value of one (1). The first byte is then followed by a 46-bit randomly generated number. A sample BSSID for an ad hoc network would be 02:12:34:56:78:9a.

Basic Service Area

The *basic service area* (BSA) defines the physical area or boundary spanned by any BSS. It is used to describe the physical area containing the members of a BSS. This area may also span other BSSs, which means that it may contain members of other BSSs.

Extended Service Set

The IEEE 802.11 standard defines an *extended service set* (ESS) as "a set of one or more interconnected basic service sets (BSSs) and integrated local area networks (LANs) that appears as a single BSS to the logical link control (LLC) layer at any station (STA) associated with one of those BSSs." The LLC is a data communication protocol sublayer in the OSI reference model. Specifically, it is a sublayer of the Data Link layer. (More about the OSI in the following section.)

An ESS is normally a union of BSSs, but the foundation for the ESS is laid as soon as a BSS is born. This means that the mere existence of a BSS implies the existence of an ESS.

The following illustration shows a sample ESS; it also shows the relative positions of some of the components of WLANs that have been discussed thus far.

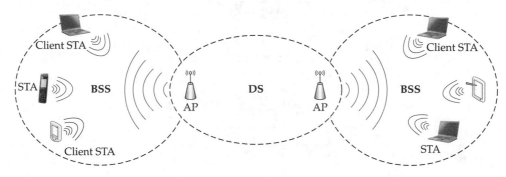

WLAN Standard (IEEE 802.11)

As you know, the IEEE 802.11 is a family of standards that governs the operations and functions of WLANs. But the standard does not define or manage absolutely every aspect of WLAN operations—it specifically concerns itself only with the functions of WLANs at the Physical (PHY) layer and Media Access Control sublayer of the OSI reference model. We examine these functions next.

Figure 6-2 shows the entire OSI model.

PHY

The Physical layer is the first layer (Layer 1) in the OSI reference model. It defines the relationship between a device and the physical communication medium.

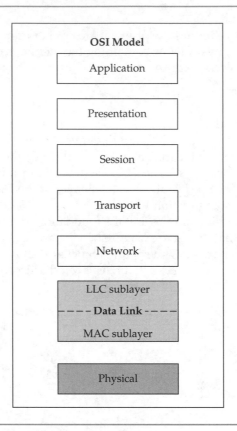

Figure 6-2. OSI reference model

For our purposes, a good example of the device is any IEEE 802.11 compliant hardware, and the physical communication medium is the airspace.

The PHY layer for IEEE 802.11 specifies the wireless signaling techniques used for transmitting and receiving information over the airwaves. Sample signaling techniques are listed in Table 6-1.

The PHY layer provides service to the IEEE 802.11 MAC sublayer, which is discussed next.

NOTE Various countries' regulatory bodies are dedicated to controlling and managing how the PHY layer of wireless devices is implemented. Wireless network administrators and other end users must therefore ensure that their wireless devices are properly configured and used so that they do not violate local laws governing the transmission and reception of RF signals. By the same token, manufacturers of wireless devices must also make sure that their devices are designed and properly labeled for use within a given region without violating the local radio spectrum use laws. See Chapter 1 for more information about radio spectrum use regulations.

Signaling Technique	Explanation
Frequency-hopping spread spectrum (FHSS)	This signaling (modulation) technique specifies use in the 2.4 GHz industrial, scientific, and medical (ISM) frequency band.
	The specific frequency range is 2.402–2.480 GHz.
	FHSS works by making the signals hop (or jump from one to another) through the allowed frequency channels in a predetermined sequence. For example, a predetermined sequence can be as simple as saying that every 5 seconds the system will hop to a new channel within the ISM frequency ranges.
	FHSS is one of the modulation techniques used in early WLAN implementations and is rarely used today. It supports data rates of 1–2 Mbit/s.
Direct-sequence spread spectrum (DSSS)	This signaling (modulation) technique specifies use in the 2.4 GHz ISM band.
	The specific frequency range is 2.400–2.497 GHz. Systems implementing this PHY can support 1 Mbit/s and 2 Mbit/s data rates.
High rate direct sequence spread spectrum (HR/DSSS)	An extension or enhancement of the DSSS PHY. It also operates in the 2.4 GHz ISM frequency bands.
	Systems implementing this PHY can provide data rates of 1, 2, 5.5, and 11 Mbit/s.
Orthogonal frequency division multiplexing (OFDM)	Specifies use in the 5 GHz Unlicensed National Information Infrastructure (UNII) frequency bands and the 2.4 GHz ISM bands. Most of the recent IEEE 802.11 standards implement this PHY and its variants.
	Compared to the other PHYs, OFDM generally supports higher data rates. Systems implementing this PHY can support 6, 9, 12, 18, 24, 36, 48, and 54 Mbit/s data rates.

(Continued)

Table 6-1. Signaling Techniques

Signaling Technique	Explanation
Extended rate PHY (ERP)	The ERP PHY specification is actually several PHY specifications in one. It provides extensions to the existing PHY specifications, such as DSSS and OFDM.
	The extensions are intended to enhance backward-compatibility and coexistence with existing PHYs. It operates in the 2.4 GHz frequency band.
	Some popular variations of ERP are:
	ERP-DSSS Provides support for systems that need to implement the ERP PHY but also need to be backward-compatible with DSSS PHY.
	ERP-OFDM Implements the OFDM PHY purely for operation, in the 2.4 GHz band. Systems implementing this PHY can support 6, 9, 12, 18, 24, 36, 48, and 54 Mbit/s data rates.
	DSSS-OFDM Provides a mixed mode (or hybrid) operation for DSSS and OFDM systems. Older DSSS systems can interpret parts of the communication (such as the header) and the newer OFDM-based systems can interpret the header and the actual data payload parts of the communication.

Table 6-1.   Signaling Techniques (*Continued*)

MAC

To maintain some semblance of sanity in data communications (and human communications), certain rules and guidelines must be established and followed. This is especially important in wireless communications because of the nature of the medium used for the communications—air or space. The rules and guidelines are specified at different layers of the OSI model.

MAC is a sublayer of the OSI's Data Link layer, or layer 2. The MAC sublayer is basically responsible for providing addressing and medium access control mechanisms that make it possible for several nodes to communicate in a network. The MAC functions are used to control and manage access to the transmission medium in a communications system.

Controlling the access of stations plugged into a wired Ethernet LAN (IEEE 802.3) is relatively simple because of the use of cables. All nodes plugged into the same network can easily sense the presence or absence of an electric current in their cables. The electric current here implies the data transmission. To coordinate access to the LAN medium, LAN stations use Carrier Sense Multiple Access with Collision Detection (CSMA/CD). The key word here is "detection."

The rules that govern the IEEE 802.11 WLANs can not easily piggyback off this same method for managing access to the shared medium used in wired LANs. And there are several reasons behind this—one reason is the absence of physical wires.

The STAs in a wireless network cannot always be guaranteed to be within earshot of each other so that they can hear (or detect) when the other STAs are transmitting. This phenomena is known as the "hidden node" problem in RF communications. Furthermore, the transmission may not even be meant or destined for the hidden node, but it still needs to use the common transmission medium shared by all the nodes.

The second reason is because the radio in most wireless LAN hardware is capable of operating in either a transmitting or receiving mode at one time—it can't usually do both at the same time. For the wireless hardware to be able to detect collisions (receive mode) while it is sending data (transmit mode), it needs to include a radio that offers such capabilities. And as has already been mentioned, this is not the case in commodity wireless LAN hardware.

So instead of attempting to detect when the medium is available for use, 802.11-based systems take a different tack by trying to avoid any type of collision in the first place. This is Carrier Sense Multiple Access with Collision Avoidance (CSMA/CA), and the key word here is "avoidance."

A popular method for implementing CSMA/CA in wireless LANs is known as the Distributed Coordination Function (DCF). The following steps show how three sample wireless STAs (STA-a, STA-b, and STA-c) might negotiate access to the wireless medium. Note that this is only one of the several methods by which CSMA/CA can be implemented.

1. STA-a needs to access the wireless medium, so it puts its radio in receiving mode to see if any other STAs are currently transmitting anything.

2. If STA-a sees that the medium is in use by STA-b, it waits until STA-b is done with its transmission. The amount of time that STA-a waits is determinate.

3. STA-a will attempt to transmit again by first checking to see if the medium is available. If so, STA-a will send out a special MAC frame called a Request To Send (RTS) frame. Also called a *control frame*, this is one of several MAC frame types (as discussed in the next section).

4. STA-c will see the special frame sent from STA-a and in turn send a Clear To Send (CTS) frame.

5. STA-a will send its message to STA-c.

6. For the communication to be considered successful, STA-c needs to send an acknowledgement confirming that it indeed received the message sent by STA-a. This message is carried in another control frame type called an Acknowledgment (ACK) frame. This is also known as *positive acknowledgement*.

7. If, for whatever reason, STA-a does not receive an ACK message from STA-c, it resends the message.

MAC Frame Types

Depending on their function, IEEE 802.11 MAC frame types can be grouped into three categories: control frames, management frames, and data frames.

Control Frames   These most basic frame types are very important for all WLAN communications and are used to support the delivery of the other (management and data) MAC frame types. All the wireless STAs must be able to see the control frames—in other words, the information in the control frames is not secret or classified in any way.

Control frames are used, for example, when a wireless STA needs to negotiate and gain access to the WLAN using CSMA/CA. Other types of control frames are the Request to Send (RTS), Clear to Send (CTS), and Acknowledgment (ACK) frames.

- **RTS**   Provides some collision avoidance mechanisms for WLANs—a way to check whether the communication medium is in use by other STAs

- **CTS**   Sent by STAs in response to the RTS frame

- **ACK**   Sent by the receiving STA to confirm successful reception of the frame in question

Management Frames   These frame types are used for management purposes on the WLAN, where they play a very important role. Management frames are used by wireless STAs whenever an STA officially wants to participate or discontinue its participation in the network and for other miscellaneous housekeeping purposes. Here are some sample management frame types:

- **Beacon frame**   A very important management MAC frame type, it performs various functions, such as time synchronization among the STAs; it also stores the value of the SSID, specifies the PHY being used, and specifies the data rates supported on the WLAN, among other things.

- **Association Request frame**   These frames are sent by the STA to request association with the AP.

- **Association Response frame**   These frames contains the AP's response to the STA regarding the STA's association request. It is either a yea or nay.

- **Reassociation Request frame**   These frames are used by STAs whenever they need to be reassociated with an AP.

- **Reassociation Response frame**   These frames are sent by the AP in response to the STAs request to reassociate with the AP.

■ **Authentication frame**   These frames are used whenever a STA needs to participate in or join a BSS. Mere association is not nearly enough—the STA needs to be authenticated to make full use of the BSS. The STA uses authentication frame types to confirm its identity.

■ **Deauthentication frame**   Authenticated STAs use these frame types to signal their intention to terminate the authenticated (secure) communications.

■ **Disassociation frame**   This frame is sent by a STA that is associated with an AP to inform the AP that it wants to discontinue the association. Note that this is not a request, and as such a response or acknowledgment or confirmation is not required from the AP.

■ **Probe Request frame**   STAs send probe request frames whenever they need to discover information about other STAs. Such information might include the capabilities of the other STA or information about the supported data rates.

■ **Probe Response frame**   This frame carries the response to probe requests.

Data Frames   These frame types are responsible for transporting the actual data payload to and from the communication end points.

Complete MAC Frame

The following illustration shows a typical 802.11 MAC frame format. Table 6-2 explains some parts of the MAC frame that might be interesting from the perspective of a wireless network administrator.

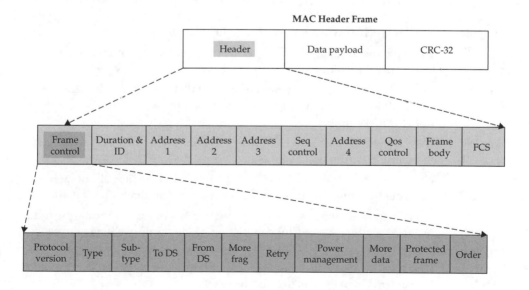

Field	Description
Frame Control	Holds important information. Divided into several fields. Sample information stored with the frame control field is related to protocol version in use, frame type (control, management, or data), fragmentation information, power management, etc.
	Can be any of the following types:
Address 1	**Source Address (SA)** A 48-bit MAC address. Serves the same purpose as the layer 2 source address used on Ethernet networks. It is the address of the STA that created the original frame to be transmitted.
Address 2	**Destination Address (DA)** A 48-bit MAC address. Serves the same purpose as the layer 2 destination address used on Ethernet networks. It is the address of the final recipient of the IEEE 802.11 frame.
Address 3	**Receiver Address (RA or Rx STA)** May or may not be the same as the final destination address (DA). May be the address of the next hop radio device on the WLAN that will forward the packet to its final destination. In an infrastructure BSS that uses an AP for all communication, RA is often the same as the BSSID—such as when the SA is relaying frames to AP for transmission.
Address 4	**Transmitter Address (TA or Tx STA)** A 48-bit MAC address. May or may not be the address of the STA that generated the original frame. TA may be an intermediate STA that transmits the frame. In an infrastructure BSS that uses an AP for all communication, TA may also correspond with the BSSID—such as when the AP is relaying frames back to the SA.
Sequence Control	Contains the fragment (of the specific fragment in question) and sequence numbers (of each frame).
Frame Body	Contains the actual data to be sent. If no data payload is to be sent, the size of this field will be 0, which is the case when the frame type is a control frame, for example.
Duration/ID	Used by wireless LAN devices to reserve or specify the time period for which RF medium will be in use.
Frame Check Sequence (FCS)	Used for checking or verifying that the frame in question did not get corrupted during transmission, using cyclic redundancy check (CRC).
Power Management	Used to indicate the power management mode of a STA. Shows the mode of the STA after a successful completion of the frame exchange sequence.
	A value of 1 indicates that the STA will be in a power-saving mode. And a value of 0 means that the STA will be in active mode. For frames transmitted by APs, this field is always 0.

Table 6-2.   Fields of an IEEE 802.11 MAC Frame

Field	Description
Protected Frame	Used to indicate when the frame body field contains information that has been processed by a cryptographic algorithm. A value of 1 means that frame body field is encrypted. A value of 1 is possible only within data frame types and some management frame types. A value of 0 is used for all other frame types.
More Data	Used as an indicator that more data is on the way. Tells a STA in power-saving mode that more data is buffered and on its way.
More Fragments	Used to indicate when more data fragments are to follow.
Retry	Receiving wireless STAs use this field to prevent processing of duplicate frames. A value of 1 in a management or data frame means that the frame is a retransmission of an earlier frame. The value is set to 0 for all other frame types.
Protocol Version	This field is 2 bits in length. As at the time of this writing, the protocol version number for the IEEE 802.11 standard is 0. All other possible values for this field are reserved for major changes in the current standard. The revision level will be incremented when a fundamental incompatibility exists between a new revision and a prior edition of the standard.
Type and Subtype	Identifies the specific function of the frame.
To DS and From DS	To Distribution System and From Distribution System.
	The meaning of these fields varies depending on the combined values in the fields. The possible values are 1 and 0, which can be combined to mean the following:
	To DS = 0 and From DS = 0
	Implies that the frame has been sent from one STA to another STA within the same IBSS (as in ad hoc) networks. Indicates a frame sent from a non-AP STA to another non-AP STA within the same BSS.
	To DS = 1 and From DS = 0
	Implies that the frame is destined for the DS or a frame being sent by a STA associated with an AP.
	To DS = 0 and From DS = 1
	When the two fields are set with these values, it means that the frame in question is exiting the DS.
	To DS = 1 and From DS = 1
	The frame uses the four-address format. This combination is possible but is currently not defined in the IEEE 802.11 standard.

Table 6-2.   Fields of an IEEE 802.11 MAC Frame

Summary

This chapter provided an inside peek into the fabric of IEEE 802.11-based networks, which can help you to understand and troubleshoot odd or complex wireless connectivity issues later on. You learned that the IEEE 802.11 family of standards guides WLAN technologies.

Some concepts and terminologies are used in everyday WLAN discussions, and you need to know these concepts and terms to understand the IEEE 802.11 standard.

The main components of the IEEE 802.11 standard are the Physical layer and the MAC sublayer of the OSI reference model. In the Physical layer of the WLAN standard are some specific PHY implementations, such as DSSS, FHSS, OFDM, and ERP.

In the MAC sublayer, you learned about the functions and different parts of a typical MAC frame.

This chapter is by no means a complete dissection of the WLAN standard as defined in IEEE 802.11. The inner workings of the current standard itself are included in a document that is almost 1300 pages long! This may seem daunting, but interested readers should rest assured that the IEEE 802.11 standard contains a lot of repetition of the same basic concepts applied to different facets of the standard.

Nevertheless, the information provided in this chapter should serve as a springboard for you as you dive into the IEEE 802.11 standard proper. Some of the subsequent chapters in this book cover certain aspects of the IEEE 802.11 standard in varying details.

CHAPTER 7 | WWANs, WMANs, and WPANs

Key Skills and Concepts

- Learn about wireless wide area networking technologies.
- Learn about wireless metropolitan area networking technologies.
- Learn about wireless personal area networking technologies.

Very few rules and standards apply to wireless wide area networks (WWANs), wireless metropolitan area networks (WMANs), and wireless personal area networks (WPANS), and we have very little control over our data or communications. This chapter offers information about these networks, including any applicable standards, incarnations, components and architecture, special features, and special interest groups (SIGs). Note that the information in this chapter is by no means a complete or thorough coverage of these technologies, but it highlights the features that you, as a wireless network administrator, might find especially interesting.

Wireless Wide Area Networks

A WAN can span a large geographical area—a country, multiple countries, and even continents. The Internet is a classic example of a WAN. A Wireless Wide Area Network (WWAN) is a class of WAN technologies that uses mostly cellular and satellite infrastructures to enable interconnectivity over a WAN via several services, such as Global System for Mobile (GSM) communication and several incarnations of GSM, the Universal Mobile Telecommunications System (UMTS), and Long Term Evolution (LTE).

 NOTE   A wireless network administrator may not have much control over the type of hardware and other implementation details of WWANs in use at his or her site, because these things are often determined by third-party service providers. However, the information about WWAN is included in this chapter because you'll find it useful to understand the technologies powering your networks. This knowledge can, for example, help when you're selecting or negotiating with service providers that will implement the actual technologies. It can also help in integrating these WWAN technologies with the existing network infrastructure that you control.

GSM Overview

GSM is a widely used digital cellular voice and data service that was initially conceived as standard that could be readily adopted by different countries and existing standards. And it has met and surpassed this initial requirement, because GSM is now adopted as a base standard in more than 80 percent of the world's mobile phone market.

GSM is considered a second generation (2G) cellular technology. The ability of wireless client devices to be truly mobile is a distinguishing feature of the 2G cellular technologies.

True mobility was made possible by the use of *cells* in mobile network designs. A cell is simply an imaginary boundary within which wireless radio frequency (RF) coverage is managed and provided by a base station. The grouping of various and consecutive cells is where the name "cellular" comes from. Various cells working in conjunction to provide expanded and seamless RF coverage to a wireless client provides mobility for the client.

Various GSM network operators own and manage their own cells, and these network operators often use sharing (also known as roaming) agreements that allow users to use their mobile devices anywhere GSM coverage is available—all at a cost, of course.

Figure 7-1 shows the relationship of the cells in cellular network and a wireless client device. Part A of the figure shows how mobility can be restricted to areas around the single cell tower. Part B shows how mobility can be enhanced when more cells are added to the scenario and mobile users can roam between cells.

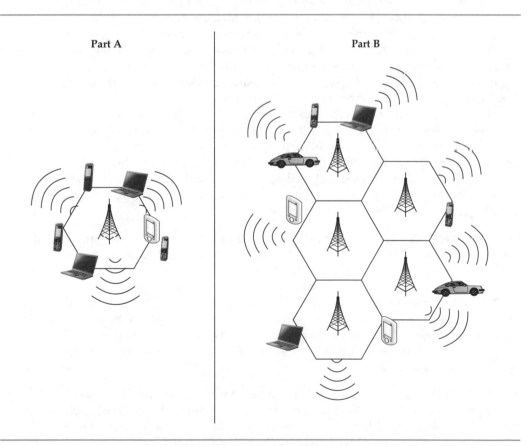

Part A   Part B

Figure 7-1.   Relationship of cells in a cellular network and wireless client device

Second-generation GSM networks operate in the 900 MHz and 1.8 GHz frequency bands in Africa, Australia, Europe, Middle East, and Asia (excluding Japan and South Korea), and parts of South America. They operate in the 850 MHz and 1.9 GHz frequency bands in North America, Latin America, and parts of South America. A more complete map of the GSM frequency bands in use in every country can be found at www.worldtimezone.com/gsm.html.

GSM Components and Architecture

GSM networks comprise several components: a mobile station (MS), a base station system (BSS), and a network switching system (NSS).

Mobile Station   The MS is the one part of the GSM network over which the user has some control (to the extent that the user is allowed to choose the color and sometimes ringtone of the wireless device). This is the part that can sometimes cost a lot of money, too. It includes any wireless client device that we intend to connect to the operator-controlled GSM network. Good examples of a MS are cell phones or a WWLAN adapter card connected to a laptop. The MS talks to the BSS (discussed next).

Every MS is uniquely identified by a International Mobile Equipment Identity (IMEI) number that is hard-coded into the GSM client device and therefore not easily transferable between devices. Among other details, the IMEI number describes the organization that registered and allocated the unique IMEI number, the model, and the vendor-assigned serial number of the GSM device.

The first time a user subscribes to the services of a GSM network operator, he or she is issued a Subscriber Identity Module (SIM) card. The SIM card is a type of smart card that stores information designed to identify a user or account uniquely on any GSM network. The SIM card is tied to the user but is portable between devices. The SIM card stores the International Mobile Subscriber Identity (IMSI), the current subscription information, user authentication data, and a simple contact database for the user (an address book). The IMSI is unique world-wide.

Base Station System   The BSS component of a GSM network is an amalgamation of components owned and managed by the network operator. It usually comprises the base station controller (BSC) and the base transceiver stations (BTS)—the radio transmitters, receivers, and antennas that serve each cell.

The BSC is the brains behind the BTS. It stores the configuration data used for managing the BTS. For example, it controls the RF power levels in the BTS, which in turn connects the cell to the NSS, from which it gets its own instructions.

The MS connects to the BSS which in turn connects to the Network Switching System, discussed next.

Network Switching System   The NSS is a central component of any GSM infrastructure. It comprises several parts that perform different complicated functions, such as call processing, subscriber-related functions, and interfacing the mobile phone network with the traditional Public Switched Telephone Network (PSTN).

GSM SIGs

The Third Generation Partnership Project (3GPP), one of the major GSM SIGs, is committed to the maintenance and development of GSM's technical specifications (www.3gpp.org). It does this by helping to unite various telecommunications standards bodies all over the world.

Another GSM SIG, the GSM Association (GSMA) represents the interests of the mobile communications industry all over the world (www.gsmworld.com). Its focus is to drive the growth of the mobile communications industry by helping to develop and create new opportunities for its members (manufacturers and suppliers of GSM-based technologies). Its objective is clearly different from that of the 3GPP, which is more concerned with standards issues.

GSM Incarnations

The original GSM standard has undergone many evolutions and revisions—too many to mention them all here. One standard, General Packet Radio Service (GPRS), is of particular interest to us as wireless network administrators. It was designed to handle data and other multimedia applications, unlike the 2G technologies, whose focus was mainly voice applications.

GPRS   GPRS is a standard used for communicating over cellular networks. It is often referred to as a two-and-a-half–generation (2.5G) cellular technology. GPRS is based on packet-switching, which offers several advantages from the perspective of a wireless mobile user interested in wireless data communications. One such advantage is cost— packet-switched services are generally cheaper than their circuit-switched counterparts, and billing is often based on the amount of actual data transferred. Packet-switched networks allow the communication medium to be shared among different users, so no expensive circuits need to be dedicated to the user. This sharing can also be a disadvantage, however, because a fixed bandwidth is not always guaranteed or available to individual users.

From the perspective of the wireless network administrator, GPRS-based devices can support the following features:

- Internet Protocol (IP) versions 4 and 6
- Wireless Application Protocol (WAP)
- Data transfer rates of 56–114 kbit/ps

GPRS can be used for data communications in a wide variety of devices, such as mobile phones, GPRS expansion cards for laptops or personal computers, and remote terminals such as point-of-sale systems.

When used for data communications, GPRS-based devices make extensive use of the notion of an Access Point Name (APN), a simple and distinct name that is meaningful only within the cellular service provider's network. The APN is distinct from access points (APs) used for purely WLAN communications, but, generally speaking, they both help to provide wireless clients access to resources available on a network.

The APN used in GPRS services specifies the external network or services that a wireless mobile device can access. These are usually one of two types: WAP APNs or Internet APNs.

The WAP APN provides access to the mobile provider's WAP content, which is often filtered and reformatted to meet WAP specifications. A sample WAP APN for a cellular service provider named Wireless WANS 'R' US 1234, Inc., could be wap.wwrs1234.com.

The Internet APN provides the mobile device access to the standard Internet-based services such as e-mail and web browsing.

A sample Internet APN for a cellular service provider named Wireless WANS 'R' US 1234, Inc., could be internet.wwrs1234.com.

UMTS Overview

The Universal Mobile Telecommunications System (UMTS) is a third-generation (3G) cellular standard that has several implementations that go by different monikers or brands. Example UMTS implementations are Freedom of Mobile Multimedia Access (FOMA) and Wideband Code Division Multiple Access (W-CDMA).

The BTS component of 2G GSM networks is replaced by a new component called *Node B* in 3G UMTS networks. The functionality provided by the BSC in 2G GSM networks is provided by a component called the Radio Network Controller (RNC) in UMTS.

UMTS SIG

The UMTS Forum (www.umts-forum.org) can be considered a UMTS SIG with very clear objectives. The Forum's objective is to help all UMTS stakeholders understand and profit from the opportunities of 3G/UMTS networks.

UMTS Incarnations

The next sections discuss some revisions and enhancements to the original UMTS standards. Some of these revisions are the High Speed Packet Access (HSPA) family of technologies, such as High Speed Downlink Packet Access (HSDPA), High Speed Uplink Packet Access (HSUPA), and HSPA Evolved (eHSPA)—aka HSPA+. The revisions are referred to as *UMTS releases*.

HSPA Overview   HSPA refers to a family of WWAN mobile technologies that provides mobile broadband access for GSM-based devices. HSPA is considered a post-3G cellular technology. It was specifically designed to offer an easy upgrade path for cellular network operators who want to deploy the post 3G technologies. HSPA-based networks operate in the 850, 1900, and 2100 MHz frequency bands.

High Speed Downlink Packet Access   HSDPA is a 3G cellular technology based on the UMTS standard. It is described in the UMTS standard Release 5. Among its other features, HSDPA offers improvements in the downlink speeds of its predecessor (HSPA), with downlink speeds of 14 Mbps.

High Speed Uplink Packet Access   HSUPA is a 3G cellular technology described in the UMTS standard Release 6. It is referred to simply as Enhanced Uplink (EUL) in some quarters.

Among its other features, HSUPA offers improvements in the uplink speeds of its predecessor (HSDPA), with uplink speeds of 5.7 Mbps.

HSUPA also supports WLAN integration.

The coverage offered by WLANs is traditionally limited and best suited for indoor use, but the coverage offered by a UMTS network can span several miles and is best suited for outdoor use. The strengths and weaknesses of these two technologies can be used to complement one another.

This particular feature makes it possible for mobile stations (such as cell phones) to use either the cellular provider's managed WWAN or the user-managed WLAN to make voice calls or for data communications. So, for example, when a user is within the reach of a WLAN or Wi-Fi signal, all voice and data communications can occur via a traditional wireless access point (WAP) or residential gateway router. The possibilities and uses of this feature are many (and, of course, can cause headaches for the wireless network administrator).

HSPA+ Evolved   HSPA+ (pronounced *HSPA plus*), often referred to as HSPA Evolved (eHSPA), is a 3G cellular technology that is described in the UMTS standard Release 7.

Among the many enhancements it offers, HSPA+ is purported to be capable of theoretical downlink speeds of 42 Mbps and uplink speeds of more than 11.5 Mbps. The Multiple Input/Multiple Output (MIMO) antenna technology is used to achieve the enhancements in HSPA+.

LTE Overview

LTE is GSM on steroids. The authors and backers of LTE describe it as an evolution of the 3G/HSPA cellular technologies. LTE is designed to be backward-compatible with GSM and HSPA technologies. "Improved spectral efficiency" is one of the strong points touted about LTE, which simply means that the technology can make more efficient use of the available radio spectrum.

LTE Special Features

LTE is purported to be capable of theoretical downlink speeds of 172 Mbps and uplink speeds of more than 50 Mbps. It will make it possible to deliver rich multimedia and bandwidth-intensive applications over long distances wirelessly.

LTE is a purely IP-based technology.

LTE SIGs

Almost everybody in the wireless community has interests in the success of LTE—from the network operators (who can make more money with less resources), to the equipment manufacturers (who can charge more for cool new hardware to support the technology), to the consumer (who pays more for the cool new hardware and for access to the better network).

LTE is a highly anticipated technology that is supposed to be a win-win situation for most stakeholders.

Wireless Metropolitan Area Networks

A MAN can span a moderately large geographical area. The scope of the area covered by a MAN is often within a city but almost certainly restricted to within a country.

A WMAN refers to wireless technologies that facilitate interconnectivity wirelessly in a metropolitan area. WMANs can be considered mid-range networks. They normally do not use the cellular network infrastructure, but instead make use of some vendor-specific technology.

In some cases, the wireless network administrator may not have much control over the type of hardware and other implementation details of WMANs in use at his or her site, because these aspects are often in the hands of third-party service providers. However the information is included here because it is useful for you to understand the technologies powering those networks. This knowledge can help, for example, when you're selecting or negotiating with service providers that will implement the actual technologies. It can also help you in integrating these WMAN technologies with an existing network infrastructure.

In the next section we'll look at one specific technology called WiMAX that is used in building WWAN networks.

 NOTE   WMAN can also be built by wirelessly connecting two or more WLANs using IEEE 802.11 standard-based equipment and protocols.

WiMAX Overview

Worldwide Interoperability for Microwave Access (WiMAX) is used in building WMAN networks. WiMAX can serve as a capable "last mile" technology, which means that it can be used to bypass the traditional cable or wired infrastructure to provide connectivity between the communications provider and the customer.

WiMAX operates in the 2–66 GHz frequency range.

WiMAX Standards

WiMAX is governed by the details specified in the IEEE 802.16 standards, where it's called WirelessMAN. Of the several revisions and versions of the standard, two are especially interesting to us here: IEEE 802.16d and IEEE 802.16e.

WiMAX Incarnations

Two incarnations of the standard are important to wireless network administrators: Fixed WiMAX and Mobile WiMAX.

- **Fixed WiMAX**   This first mainstream version of WiMAX was widely adopted. It is especially suited for point-to-multipoint (one-many) applications. Its inner workings are specified in IEEE 802.16d.

  Fixed WiMAX is purported to support wireless coverage of up to 30 miles (50 km).

- **Mobile WiMAX**   The details of the inner workings of Mobile WiMAX are governed by the IEEE 802.16e standard. It was developed to support mobile wireless clients—in other words, it was designed with the idea that users of the network will not always be in a fixed location and may be in motion while accessing the network.

  Mobile WiMAX is purported to support wireless coverage of up to 10 miles (15 km). Smart antenna technology, such as MIMO, is used in Mobile WiMAX to improve gain and provide better throughputs.

NOTE   Mobile WiMAX and the Long Term Evolution (LTE) standard provide similar functionality and are also technically similar. In fact, the two are considered competing standards, despite the fact that WiMAX is used mostly for WMAN applications and LTE is used in WWAN applications. Both are considered fourth-generation (4G) wireless network technologies.

WiMAX SIGs

Several WiMAX special interest groups exist, including the following:

- **The WiMAX Forum**   This group describes itself as an industry-led, non-profit organization dedicated to certifying and promoting compatibility and interoperability of products based on the IEEE 802.16 standard. The relationship of the WiMAX Forum to WiMAX is similar to the relationship of the Wi-Fi Alliance to WLAN technologies. The WiMAX Forum's web site is at www .wimaxforum.org/.

- **Intel Corporation**   Intel is a big proponent of WiMAX technology. This is understandable because they are possibly the biggest manufacturer of the hardware (chipsets) that implements WiMAX in the world. You can learn more about Intel's interest in WiMAX at www.intel.com/technology/wimax.

Wireless Personal Area Networks

A WPAN is used for facilitating communication between devices in a very small area. The "personal" aspect of this wireless network type came about because the devices in question are often used in the context of a personal space. And as with other communication types discussed thus far, WPAN's objective is to receive and/or send data.

WPAN can be considered a short-range network because the range and reach of this wireless network type is typically quite limited when compared with that offered by WLANs, WMANs, and WWANs. A WPAN requires little external infrastructure to operate; most WPANs are self-contained.

The IEEE 802.15 standards describe specifications for the inner workings of WPANs.

Several technologies exist for enabling WPANs, such as Bluetooth, ZigBee, Z-Wave, Infrared Data Association (IrDA), and Ultra-Wideband (UWB), to mention a few. The following sections discuss Bluetooth and ZigBee.

Bluetooth Overview

Bluetooth technology replaces the cables traditionally used for connecting numerous electronic devices.

Bluetooth uses the frequency-hopping spread spectrum (FHSS) modulation technique. Its devices operate in the 2.4–2.4835 GHz unlicensed frequency range.

Bluetooth Standards

The IEEE 802.15.1-2002 and IEEE 802.15.1-2005 standards are good examples of the few Bluetooth standards that emerged from a neutral standards body. Most of the past and future work done regarding the development of Bluetooth is accomplished by the Bluetooth SIG.

Bluetooth Components and Architecture

Bluetooth devices consist primarily of the same components that make up most wireless RF devices: transmitter and receiver components (the transceiver) and the baseband.

Protocols   The Bluetooth standard defines a group of protocols that are used to manage communications between the devices. Most commonly used Bluetooth protocols are listed in Table 7-1.

Profiles   Bluetooth devices make use of *profiles* to determine the services and protocols that are supported by the device. This is a useful but somewhat confusing feature of the technology. It is useful because it makes it easy for hardware manufacturers and Bluetooth software developers to create Bluetooth devices and applications that are very specific in scope. In other words, it helps to keep things simple and possibly bring down the costs of Bluetooth devices because they do not need to support a plethora of features.

The confusion can result on the user side because the unknowing user may assume that all Bluetooth-capable devices can be used for *any* type of communication with *any* other Bluetooth device. This is incorrect, however, because to communicate, two Bluetooth devices must be able to speak the same profile.

Some common Bluetooth profiles are listed in Table 7-2.

Bluetooth Protocol	Description
Object Exchange (OBEX)	A simple file transfer protocol that defines data objects. Bluetooth devices use this communication protocol to exchange those objects.
Bluetooth Network Encapsulation Protocol (BNEP)	Used for transporting and encapsulating Ethernet packets. It can therefore be used for transporting IPv4 and IPv6 protocols since they can in turn be encapsulated within Ethernet packets.
RFCOMM	Used for emulating serial type connections by emulating the serial cable line settings and the status of an RS-232 serial port.
Link Manager Protocol (LMP)	Used for setting up and controlling the logical transports, logical links, and physical links. Controls the communication links between devices.
Logical Link Control and Adaptation Protocol (L2CAP)	Bluetooth protocols that operate at the higher layers rely on L2CAP to provide an interface (logical channel) for communicating with the rest of the Bluetooth protocol stack.
Service Discovery Protocol (SDP)	Used to discover (by searching or browsing) and learn about the characteristics of the services offered by other Bluetooth devices.
Telephony Control Protocol (TCP)	Manages call control for voice and data calls between Bluetooth devices.
Audio/Video Control Transport Protocol (AVCTP)	Used for transporting the messages used for controlling Audio/Video (A/V) devices.
Audio/Video Distribution Transport Protocol (AVDTP)	Defines procedures for negotiating, establishing, and transmitting Audio/Video (A/V) streams.

Table 7-1.   Bluetooth Protocols

Network   Two or more Bluetooth devices communicate with one another using a *piconet*, a type of Bluetooth network that comprises one master and one or more slaves. The master is responsible for regulating all access to the Bluetooth RF channel. The bandwidth of the RF channel is shared among the participating Bluetooth devices in the piconet. Each piconet operates in its own frequency-hopping radio channel.

A piconet is exclusive, in that a maximum of only seven active slaves (and one master) are allowed to participate in the network at a time. However, other slaves are

Bluetooth Profile	Description
Generic Object Profile (GOEP)	Implements the OBEX protocol; used for file exchange—to transfer objects (files) from one Bluetooth device to another.
Dial-Up Network Profile (DUN)	Used to access the Internet and other dial-up services over Bluetooth—for example, by connecting wirelessly over Bluetooth to a mobile phone, which in turn provides the connection to the Internet.
File Transfer Profile (FTP)	Used for browsing, manipulating, and transferring objects on file systems. Provides similar functionality to TCP/IP FTP.
PAN profile	The implementation of the BNEP.
Service Discovery Application Profile (SDAP)	Specifies how an application should use the SDP to discover services on another device.
Human Interface Device Profile (HID)	Used for connecting various Bluetooth-based input devices—keyboards, mice, scanners, joysticks, etc.

Table 7-2.   Common Bluetooth Profiles

allowed to wait outside the piconet until any of the seven slaves leaves the network. These waiting slaves, in a *parked* state, can join the piconet after a vacancy has been created. Up to 255 devices can exist in an inactive, or parked, state, and the master device can bring these into active state at any time. Active member slaves are referred to as being in an *active* state.

When two or more established piconets are in proximity of each other, they form a *scatternet*, which forms as a result of the individual piconets having overlapping radio frequency coverage areas. The slaves in one piconet can participate in another piconet in the role of either a master or a slave.

Figure 7-2 shows some Bluetooth network topologies. It shows two separate piconets (Piconet-A and Piconet-B), and it also shows how the two piconets can combine to form a scatternet.

Bluetooth Incarnations

Bluetooth technology has grown and developed a lot over the years since its inception, and, as a result, a few versions are out there (as of this writing): Bluetooth 1.2, 2.0, 2.1, and 3.0.

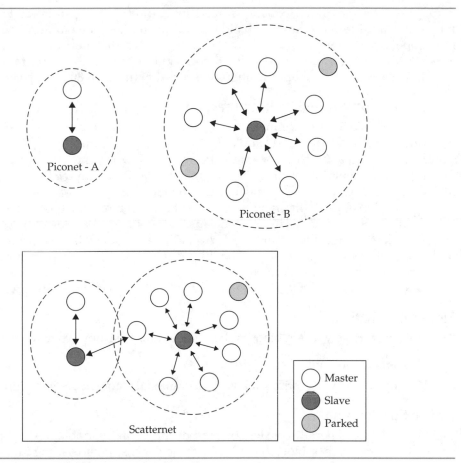

Figure 7-2.   Bluetooth network topologies

Bluetooth SIGs and Manufacturers

A prominent Bluetooth SIG is the aptly named Bluetooth SIG (www.bluetooth.org), a non-profit trade association. Among other things, the group is tasked with publishing Bluetooth specifications, qualifying Bluetooth devices, and promoting the Bluetooth trademark.

Several big-name Bluetooth hardware and software vendors abound and include Ericsson, Intel, Toshiba, Lenovo, Microsoft, Motorola, and Nokia, to name a few.

ZigBee Overview

ZigBee is especially suited for use in embedded applications that require low data rates and low power consumption. ZigBee is targeted for use in wireless monitoring and control systems and automation type applications. Its low cost, low power, and open standards–based attributes are some of its key differentiating points.

ZigBee uses the direct-sequence spread spectrum (DSSS) modulation technique. It operates in different frequency bands in various parts of the world. For example, in Europe, it works in the 868–868.8 MHz range; in North America, it works in the 902–928 MHz range; and worldwide, it works in the 2400–2483.5 MHz range.

ZigBee is based on the specifications described in the IEEE 802.15.4 standards.

ZigBee Components and Architecture

From a wireless network administrator's point of view, the components of a ZigBee network are relatively few and simple.

The players in a ZigBee network can be grouped broadly into two components: the physical components and the logical components. We start with the logical components in Table 7-3. The physical components are shown in Table 7-4.

ZigBee WPANs use a mesh network type of architecture, which is described as a "self-healing" network. This basically means that nodes (ZigBee devices) are pretty smart and knowledgeable about their surroundings and know how to work around faults and other simple kinks in the communication links between its members. ZigBee networks can support up to 64,000 nodes.

A sample wireless network made up of ZigBee devices is shown in Figure 7-3.

ZigBee Incarnations

As of this writing, few incarnations of the ZigBee suite exist, but two notable ones are the

- ZigBee, the original standard, and
- ZigBee Pro, with optimizations to accommodate more nodes in a ZigBee network.

ZigBee Special Features

ZigBee hardware generally cost less than other competing technologies that perform similar functions, such as Bluetooth. Despite the lower cost, devices using ZigBee have not yet achieved the market penetration and acceptance of the more expensive competition.

Logical Roles	Characteristics
ZigBee coordinator	Initializes a network
	Manages other network nodes
	Stores network node information
ZigBee router	Routes messages between paired nodes
ZigBee end device	Acts as a leaf node in the network

Table 7-3.   Logical Components of a ZigBee Network

Physical Component	Possible Logical Role	Characteristics
Reduced Function Device (RFD)	ZigBee end device	Used mostly for sending or receiving data; searches for available network; requests data from the network coordinator
		Cannot become a ZigBee coordinator Does not communicate directly with other RFDs
		Minimum hardware resources (e.g. memory)
		Often battery powered
Full Function Device (FFD)	ZigBee end device ZigBee coordinator	Can function as a coordinator, as a router, or as another RFD
	ZigBee router	Can be battery powered or powered directly from the mains

Table 7-4.   Physical Components of a ZigBee Network

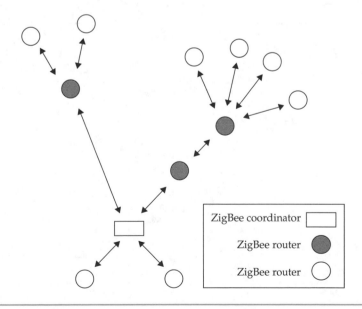

Figure 7-3.   A sample ZigBee network

ZigBee SIGs and Manufacturers

A prominent ZigBee SIG is the ZigBee Alliance, "an association of companies working together to enable reliable, cost-effective, low-power, wirelessly networked, monitoring and control products based on an open global standard," according to its web site at www.zigbee.org. Sample manufacturers that make ZigBee-based devices include Emerson, Freescale, Ember, and Philips.

Summary

Wireless Wide Area Networks (WWANs) include some specific WWAN implementations—specifically, GSM, GPRS, UMTS, and LTE.

Wireless Metropolitan Area Networks (WMANs) include WiMAX networks.

Wireless Personal Area Networks (WPANs) use Bluetooth and ZigBee implementations.

PART IV | Protocols, Services, and Security in Wireless Networks

CHAPTER 8 | Wireless Network Services and Protocols: TCP/IP

Key Skills and Concepts

■ Learn about the Transmission Control Protocol (TCP).

■ Learn about the User Datagram Protocol (UDP).

■ Learn about the Internet Protocol (IP) and IP versions 4 and 6.

■ Understand how hosts are addressed on IP networks.

At this point, you should know about various aspects of wireless networks, including the standards governing the operations of wireless networks and the various client-side and infrastructure-side hardware that implement the specifications described in the standards.

In this part of the book, we'll shift our focus slightly and consider some real-world aspects of managing a wireless network. Specifically, you'll learn about some of the protocols and services that work invisibly in the background to make wireless networks useful to the end user.

These services and protocols are essential to the proper functioning of the wireless network. They are also normally transparent to the end user—at least until one of the services becomes unavailable.

Even though the main aspects of the wireless network (Physical layer and Media Access Control sublayer) may be configured and working properly, incorrectly configured or nonfunctioning upper-layer components can make a wireless network unusable. For example, the Transmission Control Protocol and Internet Protocol (TCP/IP) suite discussed next is used extensively by the nodes in a wireless network for addressing each other, locating each other, locating other network resources, transferring packets back, and so on.

TCP/IP and UDP

Transmission Control Protocol and Internet Protocol (TCP/IP) refers to an entire suite of protocols used for network communications. Even though the TCP/IP suite consists of several other protocols, the entire suite gets its name from two very important protocols in the suite—TCP and IP.

TCP and IP are both upper layer protocols in the OSI conceptual model. Specifically, IP operates at layer 3 and TCP operates at layer 4. The two protocols complement one another and often work hand-in-hand. For this reason, they are often regarded as a single entity. The following sections, however, cover IP (version 4) and TCP individually.

Internet Protocol

IP operates at layer 3, the Network layer, of the OSI model. IP communicates directly with connected nodes in a network and connects with nodes that are not directly connected (such as nodes on other subnets, the Internet, and so on). This means that an IP packet can make its way to any other host, so long as a path exists to the destination.

Headers

Network protocols use a *header* to describe the information needed to move data from one host to the next. Packet headers, as they are typically called, consist of information that tells the protocol how to handle the packet.

Packet headers tend to be small in size; this leaves room in the packet for the actual (useful) data payload.

As each layer processes the packet, appropriate headers are removed. For example, in the case of a TCP/IP packet over a IEEE 802.11 wireless network, the driver will strip off the 802.11 Media Access Control (MAC) headers, IP will strip the IP headers, and TCP will strip the TCP headers. This will eventually leave just the data that needs to be delivered to the appropriate application.

IP helps move a packet from one host to another. Once a packet arrives at the host, no information appears in the IP header to indicate to which application the data should be delivered. IP does not provide any more features than those of a simple transport protocol because it was meant to be a foundation for other protocols. Of the protocols that use IP, not all of them need reliability, and the order of arrival of the packets is not important. Thus, it is the responsibility of higher level protocols to provide features beyond what IP provides.

IP Version 4

IPv4, as of the time of this writing, is the most prevalent version of IP in use around the world. Figure 8-1 shows the components of the header of an IP packet. We'll walk through a description of each of the fields of an IP packet header.

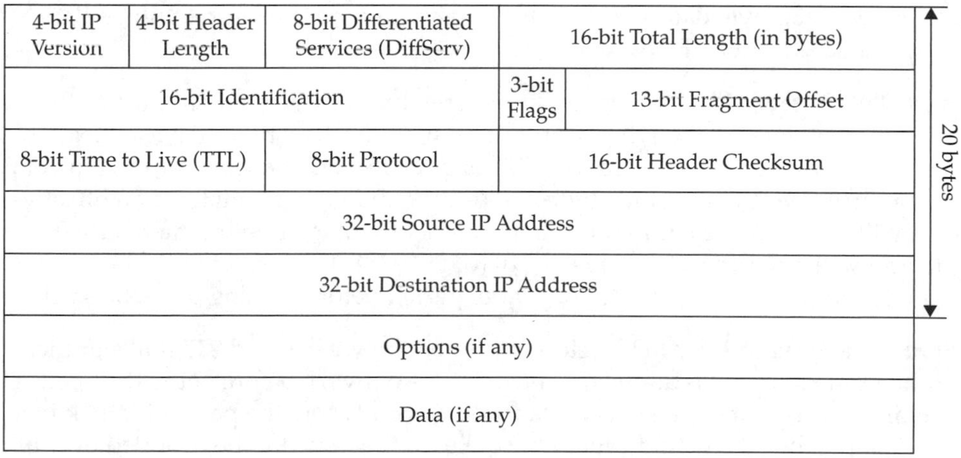

4-bit IP Version	4-bit Header Length	8-bit Differentiated Services (DiffServ)	16-bit Total Length (in bytes)	
16-bit Identification			3-bit Flags	13-bit Fragment Offset
8-bit Time to Live (TTL)		8-bit Protocol	16-bit Header Checksum	
32-bit Source IP Address				
32-bit Destination IP Address				
Options (if any)				
Data (if any)				

20 bytes

Figure 8-1. IP header

Version Field   The first value in the IP header is the version number and it shows the version of IP in use.

Length Field   The next value is the length of the IP header. The value is important to know, because optional parameters may be appended to the end of the base header. The header length tells us how many, if any, options are included. To get the byte count of the total IP header length, multiply the length number by 4. Typical IP headers will have the header length value set to 5, indicating that 20 bytes of data are included in the complete header.

Differentiated Services (DiffServ) Field   DiffServ is used for classifying, managing, and providing quality of service guarantees to network packets. It basically instructs IP stacks as to the kind of treatment to be given to a packet. (See RFC 2474, www.faqs.org/rfcs/rfc2474.html, for more details.) The use of DiffServ bits is sometimes referred to as "packet coloring," and the bits are used by networking devices for the purpose of rate shaping and prioritization.

Total Length Field   This value tells you the total length of the complete packet, including the IP and TCP headers, but not including the Ethernet headers. This value is represented in bytes. An IP packet cannot be longer than 65,535 bytes.

Identification Number Field   This field is supposed to be a unique number used by a host to identify a particular packet.

Flags Field   The flags in the IP packet tell you whether the packet is fragmented. Fragmentation occurs when an IP packet is larger than the smallest maximum transmission unit (MTU) between two hosts. MTU defines the largest packet that can be sent over a particular network. For example, Ethernet's MTU is 1500 bytes. Thus, if we have a 4000-byte (3980-byte data + 20-byte IP header) IP packet that needs to be sent over Ethernet, the packet will be fragmented into three smaller packets. The first packet might be 1500 bytes (1480-byte data + 20-byte IP header), the second packet might also be 1500 bytes (1480-byte data + 20-byte IP header), and the last packet will be 1040 bytes (1020-byte data + 20-byte IP header).

Fragmentation Offset Field   The fragment offset value indicates which part of the complete packet you are receiving. Continuing with the 4000-byte IP packet example, the first fragment will include bytes 0–1479 of data and will have an offset value of 0. The second fragment will include bytes 1480–2959 of data and will have an offset value of 185 (or 1480/8). And the third and final fragment will include fragments 2960–3999 of data and will have an offset value of 370 (or 2960/8). The receiving IP stack will reassemble these three packets into one large packet before passing it up the stack.

TTL Field   The time-to-live (TTL) field is a number between 0 and 255 that signifies how much time a packet is allowed to have on the network before being dropped. The idea behind this is that, in the event of a routing error where the packet is stuck in a "routing loop," the TTL would cause the packet to time out and be dropped eventually, thus keeping the network from becoming completely congested with looping packets.

ICMP

ICMP was designed as a means for networked hosts to communicate with one another about the state of the network. Since the data is used only by the underlying operating system and not by users, ICMP does not support the notion of port numbers, it does not require reliable delivery, and it doesn't guarantee the order of packets.

Every ICMP packet contains a type field that tells the recipient the nature of the message. The most popular type is Echo-Request, which is used by the infamous ping program. When a host receives the ICMP Echo-Request message, it responds with an ICMP Echo-Reply message. This allows the sender to confirm that the other host is up, and since we can see how long it takes the message to be sent and replied to, we get an idea of the latency of the network between the two hosts.

As each router processes the packet, the TTL value is decreased by 1. When the TTL reaches 0, the router on which this happens sends a message via Internet Control Message Protocol (ICMP), informing the sender of this (see the "ICMP" sidebar).

Protocol Field   This field tells you to which higher level protocol this packet should be delivered. Typically, this has a value for TCP, UDP, or ICMP.

Header Checksum   The last small value in this IP header is the checksum. This field holds the sum of every byte in the IP header, including any options. When a host builds an IP packet to send, it computes the IP checksum and places it into this field. The receiver can then do the same math and compare values. If the values don't match, the receiver knows that the packet was corrupted during transmission. (For example, a lightning strike creating an electrical disturbance might create packet corruption.)

Source and Destination IP Address Fields   Finally, the numbers that matter the most in an IP header are the source and destination IP addresses. These values are stored as 32-bit integers instead of the more human-readable dotted-decimal notation. For example, instead of 192.168.1.1, the value would be hexadecimal c0a80101 or decimal 3232235777.

Transmission Control Protocol

TCP operates at the transport layer (layer 4) of the OSI model. TCP provides a reliable transport for one communication *session*—that is, a single connection from a client program to a server program.

In addition to sessions, TCP also handles the ordering and retransmission of packets. If a series of packets arrive out of order, the stack will put them back into order before passing them up to the application. If a packet arrives with any kind of problem or goes missing altogether, TCP will automatically request the sender to retransmit. Finally, TCP connections are also bidirectional. This means that the client and server can send and receive data on the same connection.

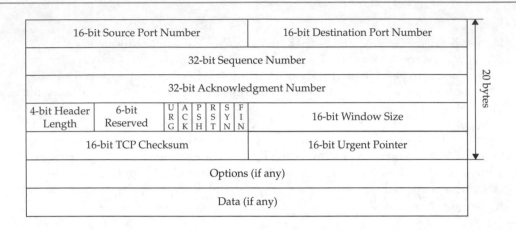

Figure 8-2.   TCP header

The TCP header is similar to the IP header in that it packs quite a bit of information into a little bit of space. You can use Figure 8-2 as a reference as you read about the various fields of a TCP packet header.

Source and Destination Port Fields   Ports are numerical identifiers used in TCP/IP-based network communications for allowing multiple process-to-process communications to occur simultaneously within a host. Port numbers are therefore used to identify processes or services running within a host. In TCP/IP, port numbers are integers between 0 and 65,535.

The first two pieces of information in a TCP header are the source and destination port numbers. Because these are only 16-bit values, their range is 0 to 65535. Typically, the source port is a value greater than 1024, since ports 1–1023 are reserved for system use on most operating systems (including Linux, Solaris, and the many variants of MS Windows). On the other hand, the destination port is typically low; most of the popular services reside there, although this is not a requirement.

Sequence and Acknowledgment Fields   The next two numbers in the TCP header are the sequence and acknowledgment numbers. These values are used by TCP to ensure that the order of packets is correct and to let the sender know which packets have been properly received. In day-to-day administrative tasks, you shouldn't have to deal with them.

Whenever the packet has the acknowledgment flag set, it can be used by the receiver to confirm how much data has been received from the sender (see "Miscellaneous Flags Field").

Header Length   Similar to IP's header length, TCP's header length tells us the header's length, including any TCP options. Whatever value appears in this field is multiplied by 4 to get the byte value.

Miscellaneous Flags Field   This field is a bit tricky. TCP uses a series of flags to indicate whether the packet is supposed to initiate a connection, contain data, or terminate a connection. The flags (in the order in which they appear) are Urgent (URG), Acknowledge (ACK), Push (PSH), Reset (RST), Synchronize (SYN), and Finish (FIN). Their meanings are as follows:

- **URG**   Implies that urgent data is in the packet and should receive priority processing.
- **ACK**   Acknowledgment of successfully received data.
- **PSH**   Request to process any received data immediately.
- **RST**   Immediately terminates the connection.
- **SYN**   Request to start a new connection.

These flags are typically used in combination with one another. For example, it is common to see PSH and ACK together. Using this combination, the sender essentially tells the receiver two things:

- Data in this packet needs to be processed.
- Acknowledges that the data packet was received successfully.

Window Size Field   This is the next field in a TCP header. TCP uses a technique called *sliding window,* which allows each side of a connection to tell the other how much buffer space it has available for dealing with connections. When a new packet arrives on a connection, the available window size decreases by the size of the packet until the operating system has a chance to move the data from TCP's input buffer to the receiving application's buffer space. Window sizes are computed on a connection-by-connection basis.

TCP Checksum Field   The TCP checksum is similar to the IP checksum in that its purpose is to give the receiver a way of verifying that the data received isn't corrupted. Unlike the IP checksum, the TCP checksum actually takes into account both the TCP header as well as the data being sent.

Urgent Pointer Field   The last piece of the TCP header, the urgent pointer, points to the offset of the octet following important data. This value is observed when the URG flag is set and tells the receiving TCP stack that some important data is present. The TCP stack is supposed to relay this information to the application so that it knows it should treat that data with special importance.

In reality, you'll be hard pressed to see a packet that uses the URG bit. Most applications have no way of knowing whether data sent to them is urgent or not, and most applications don't really care.

TCP in Action

As alluded to earlier, TCP supports the concept of a *connection*. Each connection must go through a sequence to get established; once both sides are done sending data, they must go through another sequence to close the connection.

The complete process of opening a TCP connection, sending data, and tearing down the connection is reviewed here.

Note that the information provided here has been highly simplified, and, unfortunately, because of the complex nature of TCP, it is impossible to cover every possible scenario that a TCP connection can take. However, this information should be enough to help you determine when things are going wrong at the network level.

Opening a Connection   TCP goes through a *three-way handshake* for every connection that it opens to allow both sides to send each other their state information and give each other a chance to acknowledge the receipt of that data.

1. The first packet is sent by the host that wants to open the connection with a server. Let's call this host the *client*. The client sends a TCP packet over IP and sets the TCP flag to SYN. The sequence number is the initial sequence number that the client will use for all the data it will send to the other host (the *server*).

2. The second packet is sent from the server to the client. This packet has two TCP flags set: SYN and ACK. The ACK tells the client that it has received the first (SYN) packet. This is double-checked by placing the client's sequence number in the acknowledgment field. The SYN tells the client with which sequence number the server will be sending its responses.

3. The third packet goes from the client to the server. It has only the ACK bit set in the TCP flags for the purpose of acknowledging to the server that it received its SYN. This ACK packet has the client's sequence number in the sequence number field and the server's sequence number in the acknowledgment field. This should be enough to establish a connection.

So why all the hassle to start a connection? Why can't the client just send a single packet over to the server stating, "I want to start talking—okay?" and have the server send back an "okay"? The reason is that without all three packets going back and forth, neither side is sure that the other side received the first SYN packet—and that packet is crucial to TCP's ability to provide a reliable and correctly ordered transport.

Transferring Data   With a fully established connection in place, both sides are able to send data. The data is automatically processed and made available to the actual upper layer applications that need it.

The process of the server sending some data and then getting an acknowledgment from the client can continue as long as data needs to be sent.

Closing the Connection   TCP connections have the option of ending ungracefully. That is to say, one side can tell the other "stop *now!*" Ungraceful shutdowns are accomplished with the RST (reset) flag, which the receiver does not acknowledge upon receipt. This is to keep both hosts from getting into a "RST war," where one side resets and the other side responds with a reset, thus causing a never-ending ping-pong effect.

As the first step of shutting down a connection, the side that is ready to close the connection sends a packet with the FIN bit set, indicating that it is finished. Once a host has sent a FIN packet for a particular connection, it is not allowed to send anything other than acknowledgments. This also means that even though it may be finished, the other side may still send it data. It is not until both sides send a FIN that both sides are finished. And like the SYN packet, the FIN packet must receive an acknowledgment.

And that's all there is to a graceful connection shutdown.

User Datagram Protocol

Like TCP, UDP is a popular transport layer protocol. Unlike TCP, however, it does not provide reliability. In other words, UDP won't detect lost or duplicate packets the way TCP does. UDP does, however, have its own strengths. It is a good choice for two types of traffic: short request/response transactions that fit in one packet (such as DNS) and streams of data that are better off skipping lost data and moving on (such as streaming audio and video). In the first case, UDP is better because a short request/response usually doesn't merit the overhead that TCP requires to guarantee reliability. The application is usually better off adding additional logic to retransmit on its own in the event of lost packets.

For example, when dealing with streaming data, TCP's reliability mechanisms may have undesirable effects, because it may be preferable for some lost packets to be simply skipped instead of being retransmitted. This is because human listeners/viewers are much better at handling (and much less annoyed by) short drops in audio than they are in delays.

In comparison to the header of a TCP packet, a UDP packet header is much simpler. Figure 8-3 shows a sample UDP header.

- **Source and Destination ports**   The first fields in the UDP header are conceptually the same as the TCP port numbers.

- **Length Field**   The length of the packet is specified; the field is16-bits long and specifies the length in bytes of the entire datagram, which consists of both the packet header and the actual data payload.

- **UDP checksum**   This field is used by UDP to validate that the data has arrived to its destination without corruption.

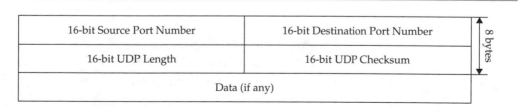

Figure 8-3.   UDP packet header

IP-Based Networks

Now that you have some of the fundamentals of TCP/IP under your belt, let's take a look at how it helps us glue networks together. This section covers hosts and networks, netmasks, static routing, and some basics in dynamic routing.

Hosts and Networks

The Internet is, of course, a large group of interconnected networks; each of these component networks has agreed to connect with some other network, and each is assigned a network address.

Traditionally, in a 32-bit IP address, the network component typically takes up 8, 16, or 24 bits to encode a class A, B, or C network, respectively. Since the remainder of the bits in the IP address are used to enumerate the host within the network, the fewer bits that are used to describe the network, the more bits are available to enumerate the hosts. For example, class A networks have 24 bits left for the host component, which means there can be upward of 16,777,214 hosts within that network. (Classes B and C have 65,534 and 254 nodes, respectively.)

NOTE   Class D and class E ranges also exist. Class D is used for multicast, and class E is reserved for experimental use.

To organize the various classes of networks, it was decided early in IP's life that the first few bits would determine to which class the address belonged. For the sake of readability, the first *octet* of the IP address specifies the class.

NOTE   An octet is 8 bits, which in the typical dotted-decimal notation of IP means the number before a dot. For example, in the IP address 192.168.1.42, the first octet is 192, the second octet is 168, and so on.

The class ranges are shown in Table 8-1.

Some special addresses are reserved for special uses. The first special address is one you are likely to be familiar with: 127.0.0.1. This is also known as the *loopback address*. It is set up on every host using IP so that it can refer to itself. It seems a bit odd to do it

Class	First Octet Range	Examples
A	0–127	1.1.1.1 and 10.0.7.9
B	128–191	128.9.8.10 and 168.0.0.78
C	192–223	192.169.0.9 and 223.1.2.25

Table 8-1.   IPv4 Address Classification

this way, but just because a system is capable of speaking IP doesn't mean it has an IP address allocated to it! On the other hand, the 127.0.0.1 address is virtually guaranteed. (If it isn't there, you know that something has probably gone wrong.)

Three other ranges are notable: Every IP in the 10.0.0.0 network, the 172.16–172.31 networks, and the 192.168 network is considered a *private IP*. These ranges are not allowed to be allocated to anyone on the Internet, and, therefore, you may use them on your internal networks.

> **NOTE**   Internal networks are networks that are behind a sort of firewall—not really directly exposed to the Internet—or that are connected to the Internet through some sort of device that is itself connected to the public Internet. Many wireless access points (WAPs) or wireless residential gateways are able to act as such a device.

Subnetting

Imagine a wireless network with a few thousand hosts on it, which is not unreasonable in a medium-sized company. Trying to tie them all together into a single large network would probably make you pull your hair out, bang your head on the wall, or possibly both. And that's just the figurative stuff.

The reasons for not keeping a network as a single large entity range from technical issues to political ones. On the technical front, there are limitations to every technology on how large a network can get before it becomes too large. Ethernet, for instance, cannot have more than 1024 hosts on a single collision domain. Realistically, having more than a dozen on an even mildly busy network will cause serious performance issues. Even migrating hosts to switches doesn't solve the entire problem, since switches, too, have limitations on how many hosts they can deal with.

Of course, you're likely to run into management issues before you hit limitations of switches; managing a single large network is difficult. Furthermore, as an organization grows, individual departments will begin compartmentalizing. Human Resources is usually the first candidate to need a secure network of its own so that nosy engineers don't peek into things they shouldn't. To support a need like that, you need to create subnetworks, a task more commonly referred to as *subnetting*.

Assuming our corporate network is 10.0.0.0, we could subnet it by setting up smaller class C networks within it, such as 10.1.1.0, 10.1.2.0, 10.1.3.0, and so on. These smaller networks would have 24-bit network components and 8-bit host components. Since the first 8 bits would be used to identify our corporate network, we could use the remaining 16 bits of the network component to specify the subnet, giving us 65,534 possible subnetworks. Of course, you don't have to use all of them!

Netmasks

The purpose of a *netmask*, often called a subnet mask, is to tell the IP stack which part of the IP address is the network and which part is the host. The netmask allows the stack to determine whether a destination IP address is on the LAN or if it needs to be sent to a router for forwarding elsewhere.

The best way to start looking at netmasks is to look at IP addresses and netmasks in their binary representations. Let's look at the 192.168.1.42 address with the netmask of 255.255.255.0:

Dotted Decimal	Binary
192.168.1.42	11000000 10101000 00000001 00101010
255.255.255.0	11111111 11111111 11111111 00000000

In this example, we want to find out what part of the IP address 192.168.1.42 is network and what part is host. Now, according to the definition of netmask, those 0 bits are part of the host. Given this definition, we see that the first three octets make up the network address and the last octet makes up the host.

In discussing network addresses with other people, you'll find it handy to be able to state the network address without explicitly providing the original IP address and netmask. Thankfully, this network address is computable, given the IP address and netmask, using a bitwise AND operation.

The way the bitwise AND operation works can be best explained by observing the behavior of two bits being ANDed together. If both bits are 1, then the result of the AND is also 1. If either bit (or both bits) is 0, the result is 0. You can see this more clearly in Table 8-2.

So computing the bitwise AND operation on 192.168.1.42 and 255.255.255.0 yields the bit pattern 11000000 10101000 00000001 00000000. Notice that the first three octets remained identical and the last octet became all zeros. In dotted-decimal notation, this reads 192.168.1.0.

NOTE   We need to give up one IP to the network address and one IP to the broadcast address. In this example, the network address is 192.168.1.0, and the broadcast address is 192.168.1.255.

Let's walk through another example. This time, we want to find the address range available to us for the network address 192.168.1.176 with a netmask of 255.255.255.240. (This type of netmask is commonly given by Internet service providers [ISPs] to business digital subscriber line [DSL] and T1 customers.)

Bit 1	Bit 2	Result of Bitwise AND
0	0	0
0	1	0
1	0	0
1	1	1

Table 8-2.   ANDing Bits

A quick breakdown of the last octet in the netmask shows us that the bit pattern for 240 is 11110000. This means that the first three octets of the network address plus 4 bits into the fourth octet are held constant (255.255.255.240 in binary is 11111111 11111111 11111111 11110000). Since the last 4 bits are variable, we know we have 16 possible addresses ($2^4 = 16$). Thus, our range goes from 192.168.1.176 to 192.168.1.192 (192 − 176 = 16).

Because it is so tedious to type out complete netmasks, most people use the abbreviated format, in which the network address is followed by a slash and the number of bits in the netmask. So the network address 192.168.1.0 with a netmask of 255.255.255.0 would be abbreviated as 192.168.1.0/24.

IPv6

IPv6 is the moniker for version 6 of the Internet Protocol. It is also referred to as IPng—Internet Protocol, the Next Generation. IPv6 offers many new features and improvements over its predecessor IPv4, including the following:

- A larger address space
- Built-in security capabilities, with network-layer encryption and authentication
- A simplified header structure
- Improved routing capabilities
- Built-in auto-configuration capabilities

IPv6 Address Format

IPv6 is able to offer an increased address space because it is 128-bits long (compared to the 32 bits for IPv4). Because an IPv6 address is 128-bits long (or 16 bytes), about 3.4×10^{38} possible addresses are available (compared to the roughly 4 billion available for IPv4).

A human being able to represent or memorize (without error) a string of digits that is 128-bits long on paper is no easy feat. Therefore, several abbreviation techniques are used to make it easier to represent or shorten an IPv6 address to make it more human-friendly. The 128 bits of an IPv6 address can be shortened by representing the digits in hexadecimal format. This effectively reduces the total length to 32 digits in hexadecimal. IPv6 addresses are written in groups of four hexadecimal numbers. The eight groups are separated by colons (:). Here's a sample IPv6 address:

```
0012:0001:0000:0000:2345:0000:0000:6789
```

The leading zeros of a section of an IPv6 address can be omitted—for example, the sample address can be shortened to this:

```
12:1:0000:0000:2345:0000:0000:6789
```

The rule also permits the preceding address to be rewritten like this:

```
12:1:0:0:2345:0:0:6789
```

One or more consecutive four-digit groups of zeros in an IPv6 address can be shortened and represented by double colon symbols (::), as long as this is done only once in the entire address. Therefore, using this rule, our sample address can be abbreviated like so:

```
12:1::2345:0:0:6789
```

Using the proviso in the preceding rule would make the following address invalid, because more than one set of double colons is in use:

```
12:1::2345::6789
```

IPv6 Address Types

Several types of IPv6 addresses can be used, and each address type has additional special address types, or scopes, that are used for different purposes. Three particularly special IPv6 address classifications are unicast, anycast, and multicast addresses. These are discussed next.

Unicast Addresses

A unicast address in IPv6 refers to a single network interface. Any packet sent to a unicast address is meant for a specific interface on a host. Examples of unicast addresses are link-local (for example, ::/128 - unspecified address, ::1/128 - loopback address, fe80::/10 - autoconfiguration addresses), global unicast, site-local, and other special addresses.

Anycast Addresses

An anycast address is assigned to multiple interfaces (possibly belonging to different hosts). Any packet sent to an anycast address will be delivered to the closest interface that shares the anycast type address—"closest" is interpreted according to the routing protocol's idea of distance, or it's simply the most easily accessible host. Hosts in a group sharing an anycast address have the same address prefix.

Multicast Addresses

An IPv6 multicast-type address is similar in functionality to an IPv4-type multicast address. A packet sent to a multicast address will be delivered to all the hosts (interfaces) that have the multicast address. The hosts (or interfaces) that make up a multicast group do not necessarily need to share the same prefix and also do not need to be connected to the same physical network.

IPv6 Backward-Compatibility

The designers of IPv6 built in backward-compatibility functionality into the version to accommodate the various hosts or sites that are not fully IPv6-compliant or ready. The support for legacy IPv4 hosts and sites is handled in several ways: compatible addresses (IPv4-compatible IPv6 address), mapped address (IPv4-mapped IPv6 address), and tunneling.

Mapped Addresses   Mapped addresses are special unicast-type addresses used by IPv6 hosts. They are used when an IPv6 host needs to send packets to an IPv4 host via a mostly IPv6 infrastructure. The format for a mapped IPv6 address is as follows: the first 80 bits are all 0's, followed by 16 bits of 1's, and ending with 32 bits of the IPv4 address.

Compatible Addresses   The compatible type of IPv6 address is used to support IPv4-only hosts or infrastructures—that is, those that do not support IPv6 in any way. It can be used when an IPv6 host wants to communicate with another IPv6 host via an IPv4 infrastructure. The first 96 bits of a compatible IPv6 address is made up of all 0's and ends with 32 bits of the IPv4 address.

Tunneling   This method is used by IPv6 hosts that need to transmit information over a legacy IPv4 infrastructure using configured tunnels. This is achieved by encapsulating an IPv6 packet in a traditional IPv4 packet and sending it via the IPv4 network.

Summary

The TCP/IP protocol is a very important part of today's wireless and wired networks and this is why an entire chapter has been devoted to it. This chapter covered the fundamentals of TCP/IP, including IP, TCP, UDP, IP addressing, subnetting, netmasks, and IP versions 4 and 6. We also examined the fields of some of the headers of the protocols that we discussed.

CHAPTER 9 | Standard Wireless Network Infrastructure Services and Protocols: DNS, DHCP

Key Skills and Concepts

■ Learn about key infrastructure services that support wireless networks.

■ Understand the Domain Name System (DNS).

■ Understand the Dynamic Host Configuration Protocol (DHCP).

Chapter 8 laid the foundation for this chapter's discussion of the standard services that exist higher up in the Open System Interconnection (OSI) conceptual model. Specifically, these protocols are implemented above the Network and Transport layers, and they rely heavily on the functionality provided by the lower layer protocols. In other words, the protocols discussed in this chapter are *encapsulated* by the lower layer protocols.

This chapter provides a cursory description of the protocols but does not go into any configuration details of the protocols themselves. The wireless network administrator is often responsible for managing and configuring these back-end services and needs to understand the roles and interrelationships of these services in the overall network infrastructure.

Domain Name System

The *Domain Name System (DNS)* provides a means of mapping numerical IP addresses (discussed in Chapter 8) into people-friendly and easy-to-remember names. This translation isn't mandatory, but it does make the network much more useful and easier to work with for humans.

This section offers some background material that will help you understand the inner workings of DNS.

Domain and Host Naming Conventions

Until now, you've most likely referenced sites by their *fully qualified domain name (FQDN)*, like this one: www.ieee802.org. Each string of characters between the dots in this FQDN is significant.

Starting from the right end and moving to the left, we have the top-level domain component, the second-level domain component, and the third-level domain component.

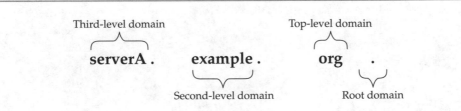

Figure 9-1.   FQDN for serverA.example.org

This is illustrated in Figure 9-1 in the FQDN for a system (serverA.example.org) and is a classic example of a FQDN. Its breakdown is discussed in detail in the following section.

The Root Domain

The DNS structure is similar to that of an inverted tree—the root of the tree is at the top and its leaves and branches are at the bottom (see Figure 9-2).

At the top of the inverted domain tree is the highest level of the DNS structure, aptly called the root domain and represented by the simple dot (.).

This is the dot that's supposed to occur after every FQDN, but it is silently assumed to be present even though it is not explicitly written. Thus, for example, the proper FQDN for www. ieee802.org is really www.ieee802.org. (notice the root period/dot at the end). And the FQDN for the popular web portal for Yahoo! is actually www.yahoo .com. (likewise).

Coincidentally (or not), this portion of the domain name space is managed by a bunch of special servers known as the *root name servers*. At the time of this writing, a total of 13 root name servers were being managed by 13 providers. (And each provider may have multiple servers distributed across the globe for various reasons, such as security and load balancing.) Also at the time of this writing, 9 of the 13 root name servers fully support IPv6 type record sets. The root name servers are named alphabetically, with names like a.root-server.net, b.root-server.net, … m.root-server.net. The role of the root name servers will be discussed in a bit.

The Top-Level Domain Names

The top-level domains (TLDs) can be regarded as the first branches that we would meet on the way down from the top of our inverted tree structure. You could say that the top-level domains provide the categorical organization of the DNS namespace. What this means in plain English is that the various branches of domain namespace have been divided into clear categories to fit different uses (examples of such uses could be geographical, functional, and so on). At the time of this writing, more than 281 TLDs existed.

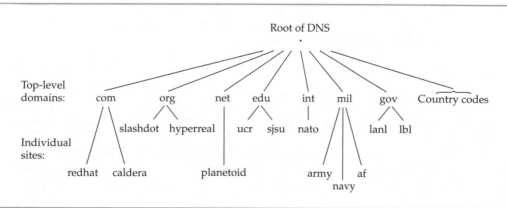

Figure 9-2.   The DNS tree, two layers deep

These TLDs can be broken down further into the generic TLDs (such as .org, .com, .net, .mil, .gov, .edu, .int, .biz), country-code TLDs (such as .us, .uk, .ng, and .ca, corresponding to the country codes for the United States, the United Kingdom, Nigeria, and Canada), and other special TLDs (such as the .arpa domain).

The TLD in our sample FQDN (serverA.example.org.) is *.org*.

The Second-Level Domain Names

The names at this level of the DNS make up the actual organizational boundary of the namespace. Companies, Internet service providers (ISPs), educational communities, non-profit groups, and individuals typically acquire unique names within this level. Here are a few examples: ieee802.org, labmanual.org, kernel.org, and caffenix.com.

The second-level domain in our sample FQDN (serverA.example.org.) is *example*.

The Third-Level Domain Names

At this level of the domain namespace, individuals and organizations that have been assigned second-level domain names can pretty much decide what to do with the third-level names. The convention, though, is to use the third-level names to reflect host names or other functional uses. It is also common for organizations to begin the subdomain definitions from here. An example of functional assignment of a third-level domain name will be the "www" in the FQDN www.yahoo.com. The "www" here can be the actual host name of the machine under the umbrella of the yahoo.com domain, or it can be an alias to a real host name.

The third-level domain name in our sample FQDN (serverA.example.org.) is *serverA*. Here it simply reflects the actual host name of our system.

By keeping DNS structured in this manner, the task of keeping track of all the hosts connected to the Internet is delegated to each site taking care of its own information.

The central repository listing of all the primary name servers, which is called the *root server,* is the only list of existing top-level domains. Obviously, a list of such a critical nature is very important, and as such it is mirrored across multiple servers and multiple geographic regions. For example, an earthquake in Japan may destroy the root server for Asia, but all the other root servers around the world can take up the slack until it comes back online. The only difference noticeable to users might be a slightly higher latency in resolving domain names.

Subdomains

While browsing the World Wide Web, you've probably come across names such as *www.support.example.org*. Such names might make you wonder which part of the whole name constitutes the *host name* component and which part makes up the *domain name* component.

Welcome to the wild and mysterious world of *subdomains*. A subdomain exhibits all the properties of a domain, except that it has delegated a subsection of the domain instead of all the hosts at a site. Using the example.org site, the subdomain for the support and help desk department of Example, Inc., would be support.example.org.

When the primary name server for the example.org domain receives a request for a host name whose FQDN ends in *support.example.org*, the primary forwards the request down to the primary name server for support.example.org. Only the primary name server for support.example.org knows all the hosts existing beneath it—hosts such as a system named *www* with the FQDN of *www.support.example.org*.

Figure 9-3 shows the relationship from the root domain to example.org and then to support.example.org. The "www" is, of course, the host name.

To make this clearer, let's follow the path of a DNS request:

1. A client wants to visit a web site called www.support.example.org.

2. The query starts with the top-level domain, *org.*, and within *org.* is *example.org*.

3. Let's say one of the authoritative DNS servers for the *example.org* domain is named *ns1.example.org*.

4. Since the host ns1 is authoritative for the example.org domain, we have to query it for all hosts (and subdomains) under it.

5. So we query it for information about the host we are interested in: *www.support .example.org*.

6. Now ns1.example.org's DNS configuration is such that for anything ending with *support.example.org*, the server must contact another authoritative server called *dns2.example.org*.

7. The request for *www.support.example.org* is then passed on to dns2.example.org, which returns the IP address for www.support.example.org—say, 192.168.1.10.

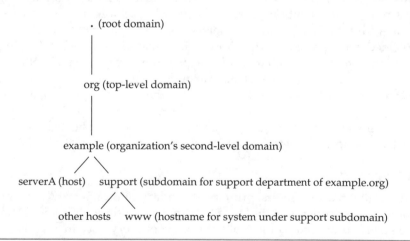

Figure 9-3.   Structure of subdomains

NOTE   Whenever a site name appears to reflect the presence of subdomains, it doesn't always imply that a subdomain is in use. Some DNS implementations allow the use of periods in the host name. Thus, from time to time, you will see periods used in host names. Whether or not a subdomain exists is handled by the configuration of the DNS server for the site. For example, www .bogus.example.org does not automatically imply that bogus.example.org is a subdomain. Rather, it may also mean that *www.bogus* is the host name for a system in the example.org domain.

Zones

The concept of DNS zones is sometimes misunderstood, and to try to clarify things here: *a DNS zone is not the same thing as a DNS domain*. The difference is subtle, but important.

Domains are designated along organizational boundaries. A single organization can be separated into smaller administrative subdomains. Each subdomain gets its own zone. All of the zones collectively form the entire domain.

For example, .example.org is a domain. Within it are the subdomains .engr.example .org, .marketing.example.org, .sales.example.org, and .admin.example.org. Each of the four subdomains has its own zone. And .example.org has some hosts within it that do not fall under any of the subdomains; thus it has a zone of its own. As a result, the example.org domain is actually composed of five zones in total.

In the simplest model, where a single domain has no subdomains, the definition of zone and domain are the same in terms of information regarding hosts, configurations, and so on.

The in-addr.arpa Domain

DNS allows resolution to work in both directions. *Forward resolution* converts names into IP addresses, and *reverse resolution* converts IP addresses back into host names. The process of reverse resolution relies on the *in-addr.arpa* domain, where *arpa* is an acronym for Address Routing and Parameters Area.

As explained in the preceding section, domain names are resolved by looking at each component from right to left, with the suffixing period indicating the root of the DNS tree. Following this logic, IP addresses must have a top-level domain as well. This domain is called the *in-addr.arpa for IPv4 type addresses*. In IPv6, the domain is called *ip6.arpa*.

Unlike FQDNs, IP addresses are resolved from left to right once they're under the in-addr.arpa domain. Each octet further narrows down the possible host names. Figure 9-4 shows a visual example of reverse resolution of the IP address 138.23.169.15.

Types of DNS Servers

DNS servers come in three flavors: primary, secondary, and caching. Another special class of name servers consists of the so-called "root name servers." Other DNS servers require the service provided by the root name servers every once in a while.

Primary Servers

Primary name servers are considered authoritative for a particular domain. An *authoritative server* is the server on which the domain's configuration files reside. Updates to the

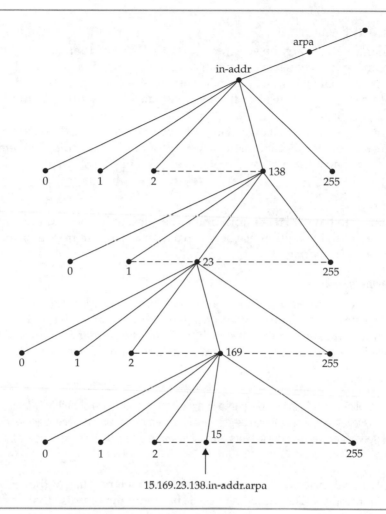

Figure 9-4.   Reverse DNS resolution of 138.23.169.15

domain's DNS tables occur on this server. A primary name server for a domain is simply a DNS server that knows about all hosts and subdomains existing under its domain.

Secondary Servers

Secondary name servers work as backups and as load distributors for the primary name servers. Primary servers know of the existence of secondaries and send them periodic updates to the name tables. When a site queries a secondary name server, the secondary responds with authority. However, because it's possible for a secondary to be queried before its primary can alert it to the latest changes, some people refer to secondaries as "not quite authoritative." Realistically speaking, you can generally

Root Name Servers

The root name servers are the very first port of call for the topmost parts of the domain namespace. These servers publish a file called the "root zone file" to other DNS servers and clients on the Internet. The root zone file describes where the authoritative servers for the DNS top-level domains (.com, .org, .ca, .ng, .hk, .uk, and so on) are located.

A root name server is just an instance of a primary name server—it delegates every request it gets to another name server. Most DNS server implementations can be configured to act like a personal root server—nothing terribly special about it.

trust secondaries to have correct information. (Besides, unless you know which is which, you cannot tell the difference between a query response from a primary and one received from a secondary.)

Caching Servers

Caching servers contain no configuration files for any particular domain. Instead, when a client host requests a caching server to resolve a name, that server will check its own local cache first. If it cannot find a match, it will find the primary server and ask it. This response is then cached. Most current DNS server implementations are capable of caching.

NOTE A DNS server can be configured to act with a specific level of authority for a particular domain. For example, a server can be primary for example.org but be secondary for domain.com. All DNS servers act as caching servers, even if they are also primary or secondary for any other domains.

Practically speaking, caching servers work quite well because of the temporal nature of DNS requests. Their effectiveness is based on the premise that, if you've asked for the IP address to example.org in the past, you are likely to do so again in the near future. Clients can tell the difference between a caching server and a primary or secondary server, because when a caching server answers a request, it answers it "nonauthoritatively."

NOTE The DNS server implementation in a majority of wireless infrastructure devices are often running in *caching-only* DNS server mode. This means that they need to be configured with the address of some other DNS server that is authoritative for the namespace in which the wireless network participates. Infrastructures devices such as access points and wireless gateways therefore forward name-resolution requests to some other regular DNS server on the network. See Figure 9-5 for an illustration of a sample wireless network with an AP configured as a caching DNS server.

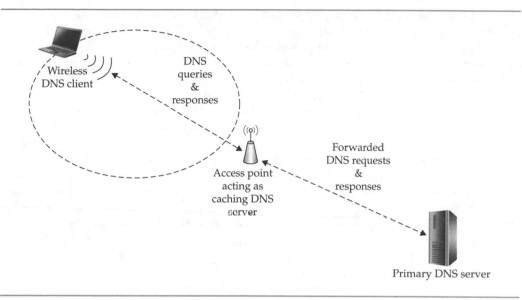

Figure 9-5. Access point acting as a caching DNS server

DNS Record Types

Name server database files store specific information that pertains to each zone that the server hosts. These database files consist mostly of record types—therefore, you need to understand the meaning and use of the common DNS record types: SOA, NS, A, PTR, CNAME, MX, TXT, and RP.

> **NOTE**   Our excursion into DNS record types assumes that the data is stored in plain-text configuration files. This is the traditional format for storing DNS records on most UNIX and UNIX-like platforms. It is possible for DNS records to be stored in other formats, such as SQL database or proprietary formats. The idea is all the same, however, and the facts about plain-text files are the same in other platforms. The bottom line is that, regardless of the format and platform on which the records are stored, a DNS record is a DNS record and will probably always be a DNS record.

SOA: Start of Authority

The SOA record starts the description of a site's DNS entries. The format of this entry is as follows (line numbers are added to the list to aid readability):

```
1  domain.name. IN SOA ns.domain.name. hostmaster.domain.name. (
2    1999080801; serial number
3    10800; refresh rate in seconds (3 hours)
4    1800; retry in seconds (30 minutes)
5    1209600; expire in seconds (2 weeks)
6    604800; minimum in seconds (1 week)
7  )
```

Line 1   Line 1 contains some important details: domain.name. is of course to be replaced with your domain name or the zone name. Notice that last period at the end of domain.name. It's supposed to be there—indeed, some DNS server implementations are extremely picky about it. The ending period is necessary for the server to differentiate relative host names from FQDNs; for example, consider the difference between serverA and serverA.example.org.

IN tells the name server that this is an Internet record. There are other types of records, but it's been years since anyone has had a need for them. You can safely ignore them.

SOA tells the name server this is the Start of Authority record.

The ns.domain.name. is the FQDN for the name server for this domain (that would be the server where this file will finally reside). Again, watch out and don't miss that trailing period.

The hostmaster.domain.name. is the e-mail address for the domain administrator. Notice the lack of an @ in this address. The @ symbol is replaced by a period. Thus, the e-mail address referred to in this example is hostmaster@domain.name. The trailing period is used here, too.

The remainder of the record starts after the opening parenthesis on line 1.

Line 2   Line 2 is the serial number. It is used to tell the name server when the file has been updated. Watch out—forgetting to increment this number when you make a change is a mistake frequently made in the process of managing DNS records. (Forgetting to put a period in the right place is another common error.)

NOTE   To maintain serial numbers in a sensible way, use the date formatted in the following order: YYYYMMDDxx. The tail-end xx represents an additional two-digit number starting with *00*, so if you make multiple updates in a day, you can still tell which is which.

Line 3   Line 3 in the list of values is the refresh rate in seconds. This value tells the secondary DNS servers how often they should query the primary server to see if the records have been updated.

Line 4   Line 4 is the retry rate in seconds. If the secondary server tries but cannot contact the primary DNS server to check for updates, the secondary server tries again after the specified number of seconds.

Line 5   Line 5 specifies the expire directive. It is intended for secondary servers that have cached the zone data. It tells these servers that if they cannot contact the primary server for an update, they should discard the value after the specified number of seconds. One to two weeks is a good value for this interval.

Line 6   The final value (the minimum) tells caching servers how long they should wait before expiring an entry if they cannot contact the primary DNS server. Five to seven days is a good guideline for this entry.

NS: Name Server

The NS record is used for specifying which name servers maintain records for this zone. If any secondary name servers exist to which you intend to transfer zones, they need to be specified here. The format of this record is as follows:

```
IN NS     ns1.domain.name.
IN NS     ns2.domain.name.
```

You can have as many backup name servers as you'd like for a domain—at least two is a good idea. Most ISPs are willing to act as secondary DNS servers if they provide connectivity for you.

A: Address Record

This is probably the most common type of record found in the wild. The A record is used for providing a mapping from host name to IP address. The format of an A address is simple:

```
Host_name     IN A      IP-Address
```

For example, an A record for the host serverB.example.org, whose IP address is 192.168.1.2, would look like this:

```
serverB     IN A      192.168.1.2
```

The equivalent of the IPv4 "A" resource record in the IPv6 world is called the "AAAA" (quad-A) resource record. For example, a quad-A record for the host serverB whose IPv6 address is 2001:DB8::2 would look like this:

```
serverB     IN AAAA     2001:DB8::2
```

Note that any host name is automatically suffixed with the domain name listed in the SOA record, unless this host name ends with a period. In the foregoing example for serverB, if the SOA record above it is for example.org, then serverB is understood to be serverB.example.org. If you were to change this to serverB.example.org (without a trailing period), the name server would understand it to be serverB.example.org .example. org.—which is probably not what you intended! So if you want to use the FQDN, be sure to suffix it with a period.

PTR: Pointer Record

The PTR record is used for performing reverse name resolution, thereby allowing someone to specify an IP address and determine the corresponding host name. The format for this record is similar to that of the A record, except the values are reversed:

```
IP-Address     IN PTR     Host_name
```

The IP-Address can take one of two forms: just the last octet of the IP address (leaving the name server to suffix it automatically with the information it has from the in-addr.arpa domain name), or the full IP address, which is suffixed with a period.

The `Host_name` must have the complete FQDN. For example, the PTR record for the host serverB would be as follows:

```
192.168.1.2.      IN PTR      serverB.example.org.
```

A PTR resource record for an IPv6 address in the ip6.arpa domain is expressed similarly to the way it is done for an IPv4 address—in reverse order. But unlike the normal IPv6 way, the address cannot be compressed or abbreviated and is expressed in the so-called reverse nibble format (four-bit aggregation). Therefore, for a PTR record for the host with the IPv6 address *2001:DB8::2*, the address will have to be expanded to its equivalent of *2001:0db8:0000:0000:0000:0000:0000:0002*.

For example, the IPv6 equivalent for a PTR record for the host serverB with the IPv6 address 2001:DB8::2 would be (single line broken here to fit on the page):

```
2.0.0.0.0.0.0.0.0.0.0.0.0.0.0.0.0.0.0.0.0.0.0.0.0.0.0.8.b.d.0.1.0.0.2. IN PTR
\ serverB.example.org.
```

MX: Mail Exchanger

The MX record is in charge of telling other sites about your zone's mail server. If a host on your network generates an outgoing mail message with its host name on it, someone returning a message would not send it back directly to that host. Instead, the replying mail server would look up the MX record for that site and send the message there instead. MX records are used, for example, when a user's desktop named pc.domain. sends a message using its PC-based mail client/reader (which cannot accept SMTP mail); it's important that the replying party have a reliable way of knowing the identity of pc.domain.name's mail server.

The format of the MX record is as follows:

```
domainname.      IN MX      weight Host_name
```

The `domainname.` is the domain name of the site (with a period at the end, of course); the `weight` is the importance of the mail server (if multiple mail servers exist, the one with the smallest number has precedence over those with larger numbers); and the `Host_name` is, of course, the name of the mail server. It is important that the `Host_name` have an A record, as well.

Here's a sample entry:

```
example.org.    IN    MX    10      smtp1
                IN    MX    20      smtp2
```

Typically, MX records occur close to the top of DNS configuration files. If a domain name is not specified, the default name is pulled from the SOA record.

CNAME: Canonical Name

CNAME records allow you to create aliases for host names. A CNAME record can be regarded as an alias. This is useful when you want to provide a highly available service with an easy-to-remember name, but still give the host a real name.

Another popular use for CNAMEs is to "create" a new server with an easy-to-remember name without having to invest in a new server at all. For example, suppose a site has a web server with a host name of *zabtsuj-content.example.org*. It can be argued that zabtsuj-content.example.org is not a very memorable or user-friendly name. So since the system is a web server, a CNAME record or alias of "www" can be created for the host. This will simply map the user-unfriendly name of zabtsuj-content.example.org to a more user-friendly name of www.example.org. This will allow all requests that go to www.example.org to be passed on transparently to the actual system that hosts the web content—zabtsuj-content.example.org.

Here's the format for the CNAME record:

```
New_host_name      IN CNAME      old_host_name
```

For example, for our sample scenario described earlier, the CNAME entry will be

```
zabtsuj-content      IN      A      192.168.1.111
www                  IN      CNAME zabtsuj-content
```

TXT and RP: The Documentation Entries

Sometimes it's useful to provide contact information as part of your database—not just as comments, but as actual records that others can query. This can be accomplished using the TXT and RP records.

A TXT record is a freeform text entry into which you can place whatever information you deem fit. Most often, you'll want to put only contact information in these records. Each TXT record must be tied to a particular host name. Here's an example:

```
serverA.example.org.      IN TXT "Contact: Admin Guy"
                          IN TXT "SysAdmin/Android"
                          IN TXT "Voice: 999-999-9999"
```

The RP record is created as an explicit container for a host's contact information. This record states who is the responsible person for the specific host. Here's an example:

```
serverB.example.org.      IN RP admin-address.example.org. example.org.
```

As useful as these records may be, they are a rarity these days, because it is perceived that they give away too much information about the site that could lead to social engineering–based attacks. You may find such records helpful in your internal DNS servers, but you should probably leave them out of anything that someone could query from the Internet.

DHCP

Manually configuring IP addresses for a handful of systems is a fairly simple task. However, manually configuring IP addresses for an entire department, building, or enterprise of heterogeneous systems can be daunting.

DHCP can assist with these tasks. A DHCP client machine can be configured to obtain its IP address from the network. When the DHCP client software is started, it broadcasts a request onto the network for an IP address. If all goes well, a DHCP server on the network will respond, issuing an address and other necessary information to complete the client's network configuration.

Such dynamic addressing is also useful for configuring mobile or temporary machines. For example, road warriors who travel from office to office can easily connect their machines to the wireless or wired local network and obtain an appropriate address for their location.

DHCP is a useful tool for dynamically configuring the addresses for large groups of machines or mobile workstations. Since DHCP is an open protocol, the architecture and platform of the server and the client are irrelevant.

NOTE   DHCP is a standard. Thus, any platform that properly implements the standard can communicate with other DHCP servers and clients regardless of their native platform.

So, for example, a wireless client running Windows, or a Linux or Macintosh OS, can be configured to use DHCP and obtain their configuration information from a proprietary but compliant wireless access point that is also acting as a DHCP server. The Windows-, Linux-, or Macintosh-based clients will not necessarily know or care that their IP configuration information is being provided by some other platform.

The Mechanics of DHCP

When a client is configured to obtain its address from the network, it asks for an address in the form of a DHCP request. A DHCP server listens for client requests. Once a request is received, it checks its local database and issues an appropriate response, which always includes the address and can include name servers, a network mask, and a default gateway. The client accepts the response from the server and configures its local settings accordingly.

The DHCP server maintains a list of addresses it can issue. Each address is issued with an associated *lease,* which dictates how long a client is allowed to use the address before it must contact the server to renew the lease. When the lease expires, the client is not expected to use the address any more. And as such, the DHCP server assumes that the address has become available and can be put back in the server's pool of addresses.

The server can be configured to issue any free address from a pool of addresses or to issue a specific address to a specific machine.

DHCP Server

The DHCP server is responsible for serving IP addresses and other relevant information upon client request. Since DHCP is broadcast-based, a server will have to be present on each subnet for which the DHCP service is to be provided.

Options

Currently, the DHCP server supports more than 60 network configuration parameters (or options) to DHCP clients.

Table 9-1 shows the most commonly used DHCP options.

DHCP Client

The DHCP client is the software component used to talk to a DHCP server described in the preceding section.

Countless platforms exists that implement the DHCP client functionality—wireless access points (WAPs); Windows, MAC, and Linux systems; wireless PDAs, network printers, and phones, to name a few.

When the client is invoked, it will attempt to obtain an address from an available DHCP server and then configure its networking configuration accordingly.

DHCP Relay

As already mentioned, a DHCP client normally operates by sending out broadcast messages. By design, routing devices do not normally forward broadcast messages. Routers stop broadcasts in their tracks. Most WAPs and wireless gateways typically

Option	Description
broadcast-address	An address on the client's subnet specified as the broadcast address
domain-name	The domain name the client should use as the local domain name when performing host lookups
domain-name-servers	The list of DNS servers for the client to use to resolve host names
host-name	The string used to identify the name of the client
routers	A list of IP addresses for routers the client is to use in order of preference
subnet-mask	The netmask the client is to use

Table 9-1.   Common DHCP Options

have some routing functionality built into them. These devices also often have built-in DHCP server functionality.

It is often desirable on any wired or wireless network to closely control the devices that can dish out IP configuration information (act as DHCP servers). This is one of the reasons why network administrators sometimes turn off the DHCP server functionality in WAPs and wireless gateways—to prevent them from also acting as DHCP servers.

One process that allows routers (wired and wireless) to forward DHCP requests from clients and DHCP responses from the server is known as *DHCP relay*.

A simplified version of how it works is detailed next and depicted in Figure 9-6.

1. A wireless node, Node-1, is configured as a DHCP client that needs to participate in an IP-based network.

2. Node-1 is connected to an infrastructure basic service set managed by an access point (AP) called AP-1.

3. AP-1 is also connected to a wired network.

4. A standalone DHCP server, called DHCPd-lan, exists on the wired network.

5. The DHCP server functionality of AP-1 is turned off.

6. Node-1 sends out broadcast message. The DHCP server DHCPd-lan does not see the broadcast message, because AP-1 has blocked it.

7. To allow Node-1 and other wireless clients to obtain IP configuration information, AP-1 will need to be configured to relay DHCP requests.

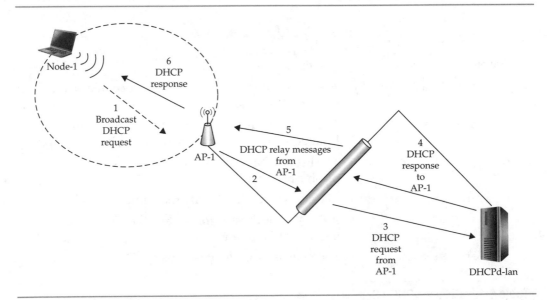

Figure 9-6.   DHCP relay

8. AP-1 gets the broadcast DHCP request. It converts the request to a unicast DHCP request packet.

9. AP-1 forwards the new unicast DHCP request packet to DHCPd-lan.

10. DHCPd-lan responds to AP-1 with the requested IP configuration information (lease)

11. AP-1 then relays the new IP configuration information it just got to Node-1.

Summary

This chapter covered the following basic information about the inner workings of some standard wireless infrastructure services.

- Name resolution
- The different DNS server types: primary, secondary, and caching servers
- Various DNS record types (for IPv4 and IPv6)
- DHCP service, which is used for providing and managing the IP configuration information for groups of client machines or mobile workstations

CHAPTER 10 | Optional Infrastructure Services and Protocols

Key Skills and Concepts

- Introduce some optional services and protocols that can enhance a wireless network infrastructure.

- Learn about the RADIUS and Diameter protocols.

- Learn when to deploy RADIUS or Diameter as a part of the wireless network infrastructure.

- Understand the functions and types of proxy servers.

- Learn when to deploy a proxy server as a part of the wireless network infrastructure.

Chapter 9 covered some standard services that complement any wireless network setup. This chapter continues that idea by discussing services that complement wireless networks. *Complement*, in this context, means that a wireless network can exist and function without these services, but these services nonetheless add value to the overall wireless network infrastructure.

As with the standard services, the services discussed in this chapter work invisibly in the background and should be transparent to the end user. The end user almost never has to directly interact with the services or even notice their presence.

RADIUS

The Remote Authentication Dial-In User Service (RADIUS) protocol is defined in the Internet Engineering Task Force (IETF) document RFC 2865 and used for managing access to resources on a network. The protocol has been around for a long time and is used in one form or another in countless applications. RADIUS does its job by providing a means for centrally managing users, which can be in the form of a user database that can also contain a list of requirements that must be met to grant access to a user.

RADIUS helps to solve some of the problems and logistics issues involved with managing large numbers of users accessing a network. RADIUS addresses these issues by taking an approach that breaks down the problems into three areas: authentication, authorization, and accounting.

The RADIUS protocol's longevity is a result of its extensibility. The protocol was designed from inception to be easily extensible so that new features can be easily layered or added to the base protocol without disturbing existing instances of the original protocol.

RADIUS Entities

Figure 10-1 shows the relative placement of RADIUS entities on a sample wireless network.

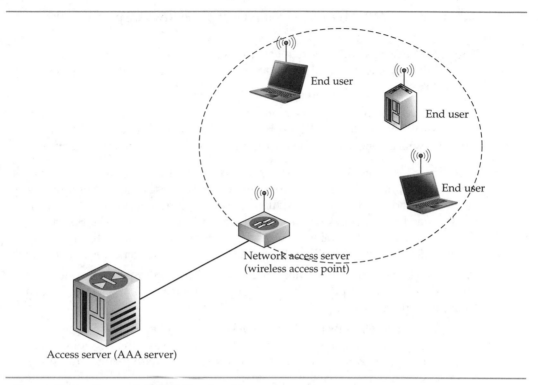

Figure 10-1. A sample RADIUS network

The following players are involved in any environment that implements RADIUS:

- **End user**   The entity that needs access to the network resource, such as a wireless station (STA) that needs to join a wireless distribution system (WDS). From the perspective of the RADIUS protocol, the end user is not necessarily the same entity as the RADIUS client. In fact, the end user is often distinct from the actual RADIUS client.

- **Network access server (NAS)**   The gate-keeping device that provides the access to the network, such as a wireless access point (AP).

   The end user connects to the NAS whenever it needs to access any network resources. The NAS in turn relies on an access server to determine whether to allow or deny access to the end user entity. The NAS is commonly known as the *RADIUS client* or the *authenticator*.

■ **Access server (AS)**   The server component of the RADIUS protocol; also known as the *RADIUS server*.

The NAS forwards end user requests to access network resources to the AS. The AS then makes the final decision about whether to grant or deny access to the end user.

You might be wondering why a book on wireless networking is discussing a dial-in user service since wireless stations (STA) obviously do not need to dial into anywhere to connect to the wireless network. The information is included for three reasons:

■ RADIUS is just a name, and the standard governing the workings of RADIUS was invented many years ago before wireless networks became as ubiquitous as they are today. The name "RADIUS" served the original purpose for which the standard was first created: to control dial-in access to networks. So, in other words, the name no longer does the standard justice, considering all the additional functionality and applications that have since been squeezed out of it.

■ Wireless STAs can dial into a network, sort of. Although the STA is not using the traditional dial-in paradigm over copper lines (cables), the STA still needs to log in to the network.

■ Large enterprise wireless (Wi-Fi) networks often rely on RADIUS as a centralized way to manage user credentials and access to the network. This helps to mitigate against some of the risks associated with using common/shared keys or passwords among users.

AAA

AAA—authentication, authorization, and accounting—can be viewed as the framework that implements the RADIUS protocol.

Authentication

Authentication is used to verify the identity of an entity. It is used, for example, *mostly* for verifying the identity of an end user. I say *mostly* because authentication can also work in the reverse direction (two-way authentication), where the entity being authenticated also requires that the authenticator (the other entity) verifies itself.

Authentication is where it all begins. Any end user or station that needs access to any network resource first needs to be authenticated. It starts when the end user presents its credentials to the NAS in one of various forms, such as username, login ID, password, hardware token, digital certificate, and so on. Authentication is considered successful when the presented credentials have been verified to be valid.

Some popular authentication methods are discussed next.

Password Authentication Protocol (PAP) This is a weak authentication method. It is weak because the user credentials are sent in clear text over network. It can be used as an authentication measure of last resort when the entities involved cannot support any other type of authentication scheme.

Challenge-Handshake Authentication Protocol (CHAP) This authentication method depends upon a shared "secret" known only to the entities involved in the schema. The secret is not sent over the link.

The process begins after a link has been established between the two entities. The authenticator sends out a challenge message to the other peer.

The peer responds to the challenge with a value calculated by using a one-way hash function. Both entities should have prior knowledge of how to apply this hash function based on the shared secret.

The authenticator also runs the same one-way hash function on the challenge and knows what the expected value should be. If the values match, the authentication is acknowledged; otherwise, the connection is terminated.

The authenticator entity periodically sends out a new challenge to the other peer during the connection lifetime to help guard against compromise.

Microsoft Challenge-Handshake Authentication Protocol (MS-CHAP) This is Microsoft's twist of the CHAP authentication method. MS-CHAP version 2 (MS-CHAP-V2) is the latest iteration of the protocol.

MS-CHAP-V2 supports mutual authentication of the peers. It also has a built-in mechanism for the authenticator to force or initiate a password change of the other peer.

Extensible Authentication Protocol (EAP) EAP is defined in the IETF documents RFC 3748 and RFC 5247 as an authentication framework that supports multiple authentication methods. EAP operates at the Data Link layer of the OSI reference model—this means that EAP is not dependent on IP.

EAP is merely a framework and not an authentication method by itself. Instead, it is a framework around which other authentication mechanisms can be designed. As implied by its name, EAP was designed with extensibility in mind, and this makes it very future-proof. As newer authentication mechanisms are developed and as vulnerabilities are found in older authentication methods, newer protocols can be designed that still use EAP as their base foundation.

EAP is used extensively in the de-facto wireless security protocols, such as Wi-Fi Protected Access (WPA and WPA2). EAP can also be used in security protocols that secure wired networks, such as IEEE 802.1X.

Authorization

The authorization component of AAA deals with permission issues. It specifies what an authorized end user can do with a given resource after the end user has been successfully authenticated. The criteria used for granting authorization can be based on end user physical location, time of day, group membership, connection type, and other factors.

Accounting

The accounting aspect of the AAA framework is used for keeping track of the end user's usage or consumption of the network resources.

Accounting can also be used for purposes of capacity planning, auditing, billing, or cost allocation, as well as to track other usage statistics, such as successful or failed user login attempts and successful or failed user authorization attempts.

NOTE   RADIUS is based on the User Datagram Protocol (UDP). The authentication component of RADIUS runs off the UDP port 1812, and the accounting component runs off the UDP port 1813.

When to Use RADIUS

So how does a wireless network administrator of great repute know when to include RADIUS as part of his or her infrastructure arsenal? The following checklist might help answer this question:

- Consider RADIUS when you need to manage and maintain a large number of users using the wireless network from a central location.

- Consider using RADIUS to complement any of the enterprise-grade wireless security solutions that can take advantage of it.

- Consider RADIUS when you need to keep track of wireless users for accounting and billing purposes.

- Consider using RADIUS when you need to enforce or define advanced policy–based features such as time-dependent access for the users of the wireless network.

- Consider using RADIUS when the wireless network is meant for the public use—such as in public wireless hotspots.

- Consider using RADIUS when a RADIUS server is already being used on the wired network, especially when the integration can be done easily, seamlessly, and in a beneficial way.

- Consider using RADIUS when you are managing a large number of disparate wireless clients that are running on different platforms, such as Linux, Windows, Macintosh, UNIX, BSD, and so on. This can help to reduce the impact of the quirks that might exist in the implementation of authentication, authorization, and encryption solutions within each vendor's platform. RADIUS is a standards-based protocol, and any platform that claims to support RADIUS must support the standard.

Popular RADIUS Implementations

Several vendors and groups have created their own implementations of the RADIUS protocol. Depending on the level of user adoption of the implementation, some of these may no longer be under active development or maintenance but may still be in use on various network deployments:

- The FreeRADIUS Project (http://freeradius.org)
- Network Policy Server (http://technet.microsoft.com/en-us/network/ bb629414.aspx), a built-in component in Microsoft Windows Server.
- BSDRadius (www.bsdradius.org)
- Cisco Secure Access Control Server (www.cisco.com)
- Juniper Networks Steel-Belted Radius series (www.juniper.net)

Diameter

It only makes sense that if we have a protocol named RADIUS, there should be a corresponding protocol named Diameter!

The Diameter protocol is the successor to the RADIUS protocol. It was conceived as a way to ease the demands being placed on the older RADIUS protocol and was deemed necessary because of the continued development and evolution of numerous access technologies that RADIUS might struggle to keep up with. Diameter was designed to be as backward-compatible with RADIUS as possible. The feature sets and capabilities of newer NAS devices have also increased substantially, thereby requiring support for similar features on the access server (AS).

Diameter is also used to provide AAA services. The inner workings of the protocol are specified in IETF RFC 3588.

The following features of the Diameter protocol distinguish it from the older RADIUS protocol:

- Diameter uses reliable transport protocols such as Transmission Control Protocol (TCP) and Stream Control Transmission Protocol (SCTP).
- It has built-in support for network or Transport layer security, such as Internet Protocol Security (IPSec) and *Transport Layer Security* (TLS).
- It offers better roaming support.
- Diameter has built-in fail-over support.
- Diameter supports automatic peer discovery and configuration by using special DNS records.
- Diameter supports Mobile IP, which allows a mobile node to change its point of attachment to the network while maintaining its original IP configuration.

You should consider using Diameter in any of the scenarios discussed in the "When to Use RADIUS" section, as well as when you need to future-proof the infrastructure components of the wireless network. Even though RADIUS is a legacy protocol, it is still very well entrenched in many IT environments, but the time will come when it will no longer be capable of handling newer technology requirements.

Diameter Entities

The following players are important in any environment that implements the Diameter protocol:

- **End user**   The entity that needs to access the network resource, such as a wireless STA that needs to participate in a WDS.

- **Diameter client**   The Diameter client generates and receives requests and responses on behalf of the user—for example, a network access server (NAS) such as a wireless access point (WAP).

- **Diameter server**   Handles AAA requests.

- **Diameter peer**   Used to describe the relationship between Diameter entities, wherein one Diameter node has a direct transport connection with another. For example, a Diameter client acting as a NAS can be peers with a Diameter server with which it is connected.

- **Diameter node**   Any entity that implements the Diameter protocol and functions either as a Diameter client, Diameter proxy, or Diameter server.

- **Diameter proxy**   An entity that helps to forward requests and responses between other Diameter nodes.

Proxy Server

Proxy servers come in all shapes, forms, and sizes, but they all have a few things in common—particularly with regard to their definition. A proxy server is software or hardware that acts as a broker or intermediary for requests from clients that need access to certain network resources. Proxy servers can help make your job much easier, can help make your users and upper management happier, and can help improve the bandwidth usage on wireless networks.

A proxy server works by receiving requests from the client, and then forwarding the client requests to the appropriate server that is hosting the resource. The proxy server also receives the responses from the server and forwards the responses back to the originating client. While all of this is going on, the proxy server may also be doing other things with the requests and responses:

- It can deny access to a given resource.

- It can speed up the entire communication between the client and server through a process called *caching*.

- It can filter or sanitize the requests and responses from clients to servers, or vice versa.

- It can offer some form of anonymity to clients by not directly exposing them to the server or to other possibly hostile external networks.

- It can reformat client or server requests or responses from one form to another.

Types of Proxy Servers

Proxy servers differ when it comes to the specific type of resource to which they are brokering access. Several types of proxy servers are based on the resource types they are proxying.

Web Proxy Server   One of the most common types of proxy servers, the web proxy server's main purpose is to proxy Web or other Hypertext Transfer Protocol (HTTP) type requests. The end objective of these types of proxy servers can be multifaceted. For example, it may function to control user access to web resources on a network, or it may be used to speed up access to web resources by caching the resources locally.

These proxy servers can also be used for changing, manipulating, or translating content from one form to another, which may be necessary when, for example, the end client cannot support or make use of the resource in its original form. A web proxy server can also be used to reformat traditional web pages to a format that is more suitable for viewing on devices such as mobile devices.

Content-filtering Proxy Server   Content-filtering proxy servers can be used for various purposes but are often used in exercising administrative control over the resources that the end user accesses.

They can also be used in restricting user access to certain web content, for forcing users to access specific content, or for enforcing compliance with network or business policies, for example.

Caching Proxy Server   Caching proxy servers are used for caching or storing copies of frequently accessed resources locally on the proxy server. This speeds up user access to these resources.

Transparent Proxy Server   These types of proxy servers are designed to work transparently in the background without any end user (or client) interaction. Because no changes need to be made to the client end, the client does not need any special support for the proxy capabilities. Transparent proxy servers are also useful in large networks where it may not be practical or feasible to reconfigure all the clients individually to use the services of the proxy server. Transparent proxy servers make it more difficult to bypass the client, which offers the administrator better control.

When to Use a Proxy Server

A wireless network administrator should consider introducing a proxy server into a wireless network if any of the following scenarios apply:

- You need to save money on network bandwidth. If your ISP charges you based on the amount of data transferred (per megabyte or per gigabyte) over their network, having an in-house proxy server will help reduce the total data transferred over your provider's network. The proxy server will try to reuse content from its local cache for your local network users instead of each user connecting to the Internet every time he or she needs the same data.

- You want to improve user experience when they're accessing external content. Content retrieval will appear faster for some data thanks to the proxy server.

- You need a way to keep track (or logs) of what the wireless network users are accessing via the network. This may be necessary for legal or administrative reasons.

- You need to control what the wireless network users are doing via the network. This may be necessary for legal or administrative reasons.

- You need to provide a simple method to authenticate users before they access web-based resources via your network.

Popular Proxy Server Implementations

Several vendors and groups have created their own proxy server products, including the following:

- Squid (www.squid-cache.org)
- Microsoft Forefront Threat Management Gateway (www.microsoft.com/ forefront/threat-management-gateway)
- Wingate (www.wingate.com)
- Tor (www.torproject.org)

Summary

You've learned about some nonstandard or optional components that can be found on wireless network infrastructures. They are considered non-standard or optional because their use is not essential to the operation of the wireless network. In general, these services or protocols are nice to have and help to make your job and life a little bit easier—and they make the network users more productive and happier as well.

CHAPTER 11 | Securing Wireless Networks: Fundamentals

Key Skills and Concepts

- Learn about cryptography and cryptographic concepts and terms.
- Learn about algorithms and ciphers.
- Review some cipher examples and implementations.
- Review the Extensible Authentication Protocol (EAP) framework.
- Review IEEE 802.11i and some considerations for deploying an IEEE 802.11i network.

Wireless-based communication systems that function over the radio frequencies have the following characteristics:

- **Invisible**   Wireless communication travels over invisible airwaves.
- **Almost boundless**   Restricting or constraining the boundaries of wireless communications can be difficult, thereby making it possible for the communications to end up in unintended locations.
- **Easy to monitor and observe using the proper equipment**   Preventing unwanted parties from monitoring or observing the communications can be difficult.

For these reasons, wireless communications have a notoriety of being insecure. Yet wireless communications are indispensable, because they offer us so much convenience. The best we can hope for is to try to find ways to manage and mitigate their undesirable characteristics.

This chapter examines some of the ways that have been developed to make wireless communications more secure. Note that making advancements in securing wireless communications, as in other IT fields, is like trying to hit a moving target. As a result, newer and better methods are constantly being developed as weaknesses or vulnerabilities are discovered in existing solutions.

Let's start by taking a brief look back into the past; then we'll work our way up to the present methods used for securing wireless communications.

Security Background

We know that communication occurs between two or more entities; the trouble with wireless communications is that, when one entity tries to communicate with another entity, a third (or more) outside entity can listen in if the proper security techniques haven't been implemented to protect the communication. In other words, it takes effort to secure wireless communications to prevent uninvited entities from listening in.

A few options are available for achieving security, including the following:

1. We don't communicate at all.

2. We communicate but restrict all communications to self.

3. We communicate but try to be careful with whom or what we communicate.

4. We communicate but try to be careful about the nature of what we communicate.

5. We communicate but with the knowledge that our communications may not be perfectly secret and accept the risks while trying to mitigate them as best as we can.

Past and current approaches to securing wireless communications tend to use a mixture of options 3, 4, and 5 to do the job.

The third option tackles the issues of communicating securely in a wireless network via means of *authentication*—either one or all the parties involved in the communications channel try to verify the others' identities.

One approach that fits the fourth option for facilitating secure wireless communications is using *cryptographic* manipulation and transformation. Cryptographic methods can be used to disguise or manipulate the communication so that it is visible or useful only to the party for which it is intended but useless to any other party.

The fourth option raises other questions and issues. For example, how can you protect or secure something that you can't see? To answer this question, we first have to understand the nature and components of the wireless "frames" being transmitted. Chapter 6 discussed the wireless Media Access Control (MAC) frame types used on IEEE 802.11 networks.

NOTE   Remember that the IEEE 802.11 standard concerns itself with the workings of the Physical layer (PHY) and the MAC sublayer of the Open Systems Interconnect (OSI) reference model. For this reason, most of the common standards-based methods and solutions for securing wireless networks are implemented at the MAC sublayer.

Let's review the MAC frame types:

- **Control frames**   These frame types are very important for all wireless communications and are used to support the delivery of the other (management and data) MAC frame types. They are the most basic frame type. It is important that the information in the control frames be visible to all the nodes in a wireless network; it is not secret in any way.

- **Management frames**   These frame types are used by wireless nodes to join or discontinue their membership in the wireless network and for other miscellaneous housekeeping purposes. Keeping the content of the management frames secret may sometimes be important.

■ **Data frames** These frame types are used for transporting the data payload. They might, for example, contain the information that we are trying to protect and transmit.

Cryptographic techniques can be used to protect information in management and data frames. The following sections discuss security concepts and techniques that can be used in securing wireless communications.

Security Services

Regardless of the solution used, any good security solution(s) employed in wireless networks should satisfy the basic needs of authentication, confidentiality, and integrity.

■ **Authentication** Ensures that the entities that need to communicate are truly who they say they are; can include authorization, which allows the entities to communicate on the network after having authenticated successfully.

■ **Confidentiality** Ensures that any information transmitted or shared between the communicating entities remains confidential or inaccessible to any unauthorized or outside entities.

■ **Integrity** Preserves the sanctity of the information communicated between the authorized entities. The contents of the communication must not be corrupted, destroyed, or altered in any way before reaching its intended destination.

Each security solution satisfies these three basic needs in varying degrees: for example, one solution might provide authentication services but is insufficient at providing confidentiality services; another solution may provide confidentiality and integrity services but cannot provide authentication services. In other words, we would find that in the real world, no security solution is perfect.

Cryptographic Concepts and Terms

Cryptography can be defined as the act or art of writing in secret characters. In technical jargon, it refers to the science and study of encrypting and decrypting information, identity verification and authorization, digital signatures, integrity checking, and secure computation. The following terms related to cryptography will aid in your understanding of later sections of this chapter.

Plain-text

Plain-text describes the unencrypted payload; it refers to the original bits and bytes as they exist before they undergo any type of cryptographic transformation.

NOTE The term "plain-text" is somewhat of a misnomer, because the word "text" here does not necessarily refer to regular text—it can be any combination of bits and bytes.

If privacy or secrecy were not of concern in wireless communication systems, the steps involved in sending and receiving data would be very simple and straightforward: We'd simply start with plain-text at the sender end, that plain-text would be transmitted as is, and then we'd end up with plain-text at the receiver end. However, because secure communications are required between senders and receivers, encryption is added to the mix. We start off with plain-text at the sender end, the plain-text then undergoes some cryptographic transformation and manipulation (becomes *cipher-text*), the cipher-text is transmitted, the receiver gets the cipher-text and passes it through another cryptographic transformation or manipulation, and, finally, if the transformation process is successful, the original plain-text is derived and viewed by the receiver.

NOTE "Clear text" is another name for plain-text. It also means that the data is unencrypted.

Encryption and Decryption

Encryption is the transformation of information from one form (plain-text) to another (cipher-text). Encryption makes cipher-text nearly impossible to decipher without the appropriate knowledge, or *key*.

Decryption, the opposite of encryption, is the transformation of encrypted information (cipher-text) back into an intelligible form (plain-text).

Key

A *key* in cryptology is similar to a key we use for locking and unlocking things in everyday life. In cryptography, keys are the bits and bytes used in the process of encryption and decryption. In this case, a key is a very large number that has special mathematical properties. Breaking into an encryption scheme depends on knowledge of the key or the ability to discover the key. The larger the key, the more difficult it is to discover.

Low-grade encryption uses 56 bits—this means 2^{56} possible keys. The following might help provide a sense of scale:

- 2^{32} is equal to 4,294,967,296 (more than 4 billion).
- 2^{48} is equal to 281,474,976,710,656.
- 2^{56} is equal to 72,057,594,037,927,936 (more than 72,057 trillion).

While this seems like a significant number of possibilities, present-day computers have enough processing power to make discovery a possibility and a cause for real concern—especially when low-grade encryption is in use.

Keyspace

The *keyspace* is related to the key used for encryption and decryption. The keyspace refers to the range of possible values that can be used in the key. The wider the range of the keyspace, the more difficult it can be to break or compromise the encryption.

Keystream

A *keystream* is a stream of random or pseudo-random data that is combined with the plain-text to produce cipher-text. Different mathematical or logic operations can be performed on the keystream in combination with the plain-text to generate the cipher-text. The keystream is not necessarily related to the key or the safeguarded information in any way.

The following logic operations can be performed on the keystream and plain-text (among others):

- Conjunction (AND)
- Disjunction (OR)
- Negation (NOT)
- Not both (NAND)
- Neither Nor (NOR)
- Exclusive OR (XOR)

 NOTE Pseudo-random numbers are a series of numbers that are similar to random numbers but cannot be defined as being completely random, because they are generated from a relatively small set of initial values. Their scope is limited or finite.

Exclusive OR (XOR)

An XOR is binary logic operation: it requires two inputs or operands. The operands are the values of propositions (a statement in which something is confirmed as true or denied as false). An XOR operation between two operands returns a true value if only one of its conditions is true.

An XOR is used as the basis for encryption in Wired Equivalent Privacy (WEP), which is discussed in a bit later in this chapter.

The following examples show the results of performing an XOR operation on two numbers—1 and 0:

- 1 XOR 1 = 0
- 1 XOR 0 = 1
- 0 XOR 1 = 1
- 0 XOR 0 = 0

Where 0 = False and 1 = True.

Algorithm

An *algorithm* is an established computational approach for solving a problem in a finite number of steps that are easily duplicable in the same manner given the same set of parameters. Technically, the algorithm must generate a result after a finite number of steps. In the world of cryptography, there are currently two classes of algorithms: *symmetric* and *asymmetric* algorithms.

Symmetric Encryption Algorithms

Symmetric algorithms often use the same key for encryption and decryption. In symmetric encrypted algorithms, the key size is directly correlated with the strength of the encryption, so a large key size used with a good algorithm will be more difficult to break than a small key size. Symmetric algorithms are best suited for providing confidentiality requirements of a security system.

This type of algorithm is also referred to as a "secret key–based algorithm" because its mode of operation depends on a secret—the "key"—that is shared between the entities that need to communicate securely. It is therefore important that the secret, or key, is well protected at all times. The sharing of this key is also one of the main weaknesses of symmetric encryption algorithms. The other weakness of the symmetric algorithms is in the logistics of how the key is distributed to the parties that need it.

Symmetric algorithms are generally computationally less demanding when compared with asymmetric algorithm (discussed next) and as such tend to execute faster.

Following are examples of symmetric algorithms:

- Advanced Encryption Standard (AES)
- Triple Data Encryption Algorithm (3DES)
- Blowfish
- Carlisle Adams and Stafford Tavares v5 (CAST5)
- International Data Encryption Algorithm (IDEA)
- Twofish

Asymmetric Encryption Algorithms

This class of algorithms employs a different key for encryption and decryption. Furthermore, the decryption key cannot be derived from the encryption key. This type of algorithm is also referred to as a "public-private key–based algorithm." The public portion of the name stems from the fact that the public key can be known by anybody (it's not a secret), but the private key portion should be a secret. The public and private keys are, however, mathematically related.

Algorithms that use asymmetric encryption are computationally more demanding than symmetric algorithms and as such are much slower in execution. The ease of distribution and management of the keys used for encryption and decryption are better than symmetric-based solutions. Some examples of asymmetric algorithms include Diffie-Hellman, ElGamal, and Merkle-Hellman Knapsack.

Public-Private Key Cryptography   Asymmetric encryption and decryption algorithms rely on a technology called *public-key cryptography*, which works similarly to a safe deposit box at the bank: you need two keys to open the box, or at least multiple layers of security checks must be used. With public-key cryptography, you need two mathematical keys: a public one and a private one. Your public key can be published on a public web page, printed on a T-shirt, or posted on a billboard in the busiest part of town. Anyone who asks for it can have a copy. On the other hand, your private key must be protected to the best of your ability. This piece of information makes the data you want to encrypt truly secure.

Every public key/private key combination is unique. So the basic principle of public-private key cryptography is that data encrypted with the public key can be decrypted using only the private key. Furthermore, encrypting with the private key can serve as a digital signature.

The actual process of encrypting data and sending it from one person to the next requires several steps. We'll use the popular "Alice and Bob analogy" to go through the process one step at a time as they both try to communicate with one another in a secure manner. Figures 11-1 through 11-5 illustrate a simplified version of the actual process.

Notice that at no point was the secret (private) key sent over the network. Once the data was encrypted with Bob's public key and signed with Alice's private key, the only pair of keys that could decrypt and verify it were Bob's private key and Alice's public key. Thus, if someone intercepted the data in the middle of the transmission, he or she wouldn't be able to decrypt the data without the proper private keys.

To make things even more secure, asymmetric encryption algorithms encourage implementations to change the session key regularly through a randomly generated, symmetric key for encrypting the communication between the sender and the receiver. Using sessions keys ensures that the data stream gets encrypted differently every few minutes. So even if someone happened to figure out the key for a transmission, that miracle would be valid for only a few minutes, until the keys changed again.

Cipher

The *cipher* is any method that is used for encryption and decryption. The meaning of the words "cipher" and "algorithm" can often be interchanged, since they both mean

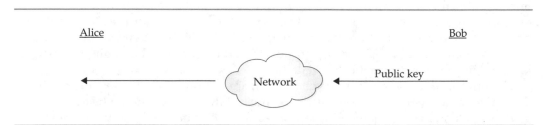

Figure 11-1.   Alice fetches Bob's public key.

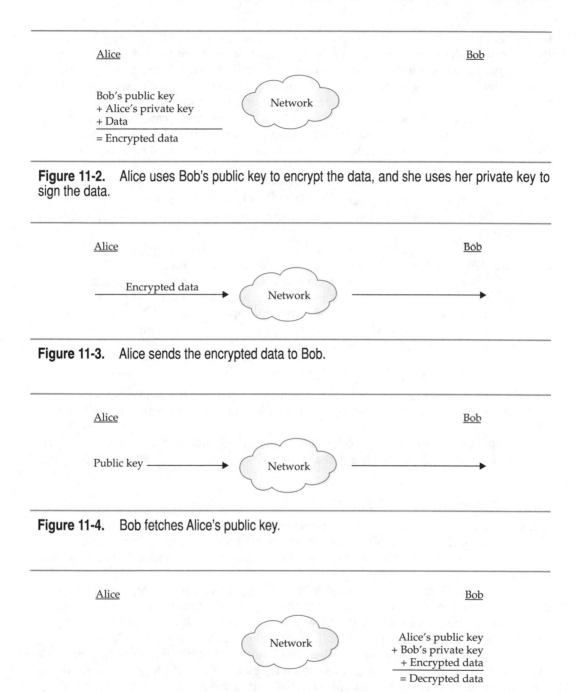

Figure 11-2. Alice uses Bob's public key to encrypt the data, and she uses her private key to sign the data.

Figure 11-3. Alice sends the encrypted data to Bob.

Figure 11-4. Bob fetches Alice's public key.

Figure 11-5. Bob uses his private key to decrypt the data, and he uses Alice's public key to verify the data.

the same thing. A cipher is the algorithm or function that creates the encryption and the decryption. These algorithms can range in complexity from the simple and elegant to the incredibly complicated and convoluted. Note that complicated ciphers do not always yield the strongest or most secure encryption. In fact, the inner workings of a good cipher system should lend itself to scrutiny and thus improvement.

Depending on their mode of implementation, ciphers can be categorized in several ways—for example, concealment ciphers versus running key ciphers, substitution ciphers versus transposition ciphers, or stream ciphers versus block ciphers.

Concealment Ciphers vs. Running Key Ciphers

Both of these ciphers types are early or classical methods for producing cipher-text.

Concealment ciphers work by concealing or hiding the plain-text message within another message.

Running-key ciphers rely on a simple square table of alphabets called a "tabula recta" and a chunk of text from a previously agreed-upon source called a "polyalphabetic source" (such as a book). The tabula recta is a table made up of rows of the alphabet, with each row computed by shifting the preceding letter to the left. The plain-text is substituted with chunks of text from the polyalphabetic source, and the cipher-text is then derived by the value in tabula recta. The "key" here is the polyalphabetic source. Table 11-1 shows a truncated tabula recta.

Substitution Ciphers vs. Transposition Ciphers

Substitution ciphers work by replacing/substituting parts or the whole of the plain-text with something else. A key is used to predetermine how the substitution should take place.

Transposition ciphers, on the other hand, do not rely on substitution; instead, parts of the plain-text are moved or juggled around to hide the meaning of the original plain-text.

Stream Ciphers vs. Block Ciphers

Stream ciphers are a special class of ciphers in which the encryption and decryption algorithm is applied to the individual bits or bytes of the plain-text. The algorithm works by combining the plain-text bits or bytes with a pseudo-random bit stream, one bit or byte at a time.

Stream ciphers are especially well suited for encrypting and decrypting the type of data that is used in network communication systems—data in transit. Some examples of a stream cipher algorithm are the RC4 cipher and the A5 algorithm that is used in cellular-based Global System for Mobile (GSM) communications.

Block ciphers are another special class of ciphers that perform their magic on blocks of plain-text instead of individual bits. When necessary, the plain-text can be divided into blocks and the algorithm is applied to the individual blocks.

	A	B	C	D	E	F	G	H	I	J	K	L	M	N	O	P	Q	R	S	T	U	V	W	X	Y	Z
A	A	B	C	D	E	F	G	H	I	J	K	L	M	N	O	P	Q	R	S	T	U	V	W	X	Y	Z
B	B	C	D	E	F	G	H	I	J	K	L	M	N	O	P	Q	R	S	T	U	V	W	X	Y	Z	A
C	C	D	E	F	G	H	I	J	K	L	M	N	O	P	Q	R	S	T	U	V	W	X	Y	Z	A	B
D	D	E	F	G	H	I	J	K	L	M	N	O	P	Q	R	S	T	U	V	W	X	Y	Z	A	B	C

Table 11-1. Tabula Recta

Block ciphers define different "operating modes." These operating modes serve as a sort of blueprint for the algorithm to perform the actual encryption or decryption. Following are some popular operating modes for block ciphers:

- **Electronic codebook mode (ECB)**   This mode is quite simple and is also prone to several weaknesses. It relies on the use of a fixed "code" book for encrypting data blocks. The fact that a given plain-text and key combination will always yield the same cipher-text is one of its weaknesses.

- **Cipher-block chaining mode (CBC)**   This mode operates by using (or chaining) the cipher-text extracted from the preceding block to encrypt the next block of data. This is probably the most popular and widely used mode of operation for block ciphers.

- **Counter mode (CM)**   This mode operates by using an initialization vector (IV) counter that increments for every block of plain-text to be converted to cipher-text. Block ciphers that operate in this mode are generally very fast in executing.

- **Output feedback mode (OFB)**   This mode is best suited for encrypting smaller chunks of plain-text at a time, because it can emulate the behavior of stream ciphers.

Cipher Examples

Following are several examples of ciphers:

DEA (DES)   A very long time ago, the National Institute of Standards and Technology (NIST) realized that sensitive data needed to be protected in a standardized way. The Data Encryption Standard (DES) was born, and this standard was implemented with the Data Encryption Algorithm (DEA).

Major parts of the inner workings of DEA were invented by the IBM corporation; it was originally known as the Lucifer cipher within IBM. NIST accepted, modified, and re-branded the Lucifer cipher.

DEA is a symmetric block cipher. This means that it works by performing a series of transpositions and substitutions on blocks of the plain-text. Its symmetric aspect means that it uses the same key for encryption and decryption. The key size used in DEA is 64 bits long.

 NOTE   IBM's original Lucifer cipher from which DEA was derived used 128-bit-long keys. For whatever reasons, NIST chose to downgrade the key length to 64 bits.

DES and its associated algorithm were in use for a long time in many applications, and eventually NIST deployed a newer and stronger algorithm. NIST determined that it was time to retire the older DEA partly because of the vastly increased computing capabilities of the then-current computer systems. In fact, the second shoe finally dropped when the efficacy of DES was compromised in 1998.

3DES (AES)   NIST called for the development and submission for a newer, better, and stronger method of protecting sensitive information after it was decided that the aging DEA was no longer sufficient, and the Advanced Encryption Standard (AES) was born. AES was implemented with the 3DES (pronounced Triple-DES), also known as the Rijndael Algorithm, which is a play of the names of its two inventors, Joan Daemen and Vincent Rijmen.

Like its predecessor, 3DES is also a symmetric block cipher, but it supports larger block sizes than DES. Specifically, it supports plain-text block sizes of 128, 192, and 256 bits, as well as corresponding key lengths of 128, 192, and 256 bits for encryption.

Generating cipher-text from plain-text using 3DES is generally a more computationally demanding process when compared to DES. This means that more powerful hardware is required when working with 3DES.

RC4   Rivest Cipher 4 (RC4), named after its inventor, Ron Rivest, is a popular and widely used cipher. It uses symmetric keys and is a stream cipher. The key size or length used in RC4 is variable but should be between 1 and 255 bytes.

Being a stream-based cipher, RC4 is quite fast in its execution and not too computationally demanding. Because it is stream-based, encryption and decryption operate on the individual bits (or bytes) of the plain-text.

Cipher Implementations

So far we've reviewed algorithms, cipher-text, plain-text, and ciphers, and you might be wondering of what practical use any of these things are to a wireless network administrator.

In the following sections, we will look at the practical implementations of some of the cryptographic concepts discussed thus far—as they pertain to wireless network administration.

Wired Equivalent Privacy Protocol

When the IEEE 802.11 standard was being developed, it was a necessity that some method be required for securing the communications that use radio frequency (RF) as a medium. The proposed and implemented solution was the Wired Equivalent Privacy (WEP) protocol. The phrase "wired equivalent" refers to the intention and goal for WEP, which was to secure communications in a wireless network in ways that are equivalent to those achieved in wired networks.

Under the hood of WEP is the RC4 cipher. WEP is a living and practical example of a textbook cipher (RC4).

The following steps show a simplified version of how WEP works:

1. The entities (nodes) that want to communicate securely in a wireless distribution systems agree to use WEP to secure their communications.

2. Since WEP is based on RC4, which uses symmetric keys, a key (common secret) is therefore needed. The nodes choose and agree on a common secret key to be used.

3. The plain-text message to be sent over the wireless medium is created by the node.

4. The plain-text data is passed through a hashing algorithm to obtain a checksum.

5. The checksum is appended to the plain-text data.

6. A 24-bits-long initialization vector (IV) is generated and appended to the secret key. This results in a stream of data that is generated and run through the RC4 pseudo-random number generator to form a keystream that is equal in length to the original plain-text/checksum from step 5.

7. An XOR operation is performed on the resulting keystream and the plain-text/ checksum combination obtained from step 5.

8. The result of the XOR operation is the cipher-text or encrypted message.

9. The cipher-text is transmitted over the wireless medium to the receiving node, which then proceeds to decrypt the data.

WEP is considered a deprecated method for securing wireless networks. WEP is easily broken and should not be relied on as a method for securing wireless communications. WEP is also vulnerable to brute-force attacks, dictionary attacks, reinjection attacks, and attacks that take advantage of its use of IV.

Wi-Fi Protected Access

WPA is the successor to the WEP protocol. After the weaknesses in the WEP implementations were discovered, WPA served as a stop-gap measure for securing wireless networks while the IEEE 802.11 standards committee developed a more robust and permanent solution. The inner workings of WPA were specified in the early versions of the IEEE 802.11i standard.

Two flavors of WPA exist: Temporal Key Integrity Protocol WPA (TKIP/WPA) and Advanced Encryption Standard WPA (AES/WPA).

TKIP/WPA

TKIP is one of the components of the IEEE 802.11i standard that attempts to address some of the vulnerabilities discovered in the WEP protocol. TKIP is used for generating random values that can be used along with the encryption algorithm to produce better encryption relative to WEP. TKIP also addresses an important aspect of a cryptographic solution that was sorely missing in WEP—*key management*. Specifically, WEP was incapable of dynamic key management. Proper key management schemes make sure that the cryptographic keys are kept safe from unauthorized parties.

A guiding principle behind the development of TKIP was to provide a solution that would not be too much of a radical departure from the existing WEP deployments. TKIP was designed so that original equipment manufacturers (OEM) of networking hardware could easily apply the new changes and improvements via software or firmware updates, without requiring the numerous/existing WEP deployments to make major changes to their current infrastructure.

So the "marketing name" for TKIP is WPA, and TKIP/WPA is endorsed by the Wi-Fi Alliance, which comes in the form of a certification program. The Alliance guarantees interoperability and compliance in products that have passed its certification requirements.

Underneath TKIP is still the RC4 algorithm—albeit a new and improved implementation. Exploitable vulnerabilities have been discovered in TKIP/WPA solution.

TIP   The version of WPA that is targeted and well suited for the home user or SOHO market is called WPA Personal, WPA Pre-shared Key (WPA-PSK). And the version targeted for the enterprise market is called WPA Enterprise.

The pre-shared key is a pass-phrase—a sequence of 8–63 ASCII-encoded characters.

Wi-Fi Protected Access 2 (WPA2)

WPA2 is based on the ratified version of the IEEE 802.11i standard. This makes it quite different from plain WPA, which was a stop-gap measure based on draft versions of the IEEE 802.11i standard. WPA2 is the Wi-Fi Alliance's moniker for its implementation of the final IEEE 802.11i standard.

Key Management

The keys are the jewels in any crypto system because they are used in the encryption and decryption process. Key management, an important aspect of any cryptographic system, includes the following:

- A mechanism to protect the keys

- A mechanism for the communicating parties to use in agreeing on the secret keys to use

- A mechanism for ensuring that unauthorized parties cannot use the keys

- A mechanism for ensuring that the keys are used for their intended purpose

- A secure method for distributing or transporting the secret keys to the parties that need them

- Automatic and periodic generation of new keys

WPA2 has several components, discussed in the following sections, that can work individually or collectively to provide what the IEEE 802.11i subcommittee calls a "robust security network."

CCMP/AES

CCMP is an awkward mouthful of an acronym that stands for Counter Mode with Cipher Block Chaining Message Authentication. CCMP provides authentication, confidentiality, and integrity checking services to any cryptographic system in which it is used. Under the hood of CCMP is the AES algorithm.

Let's try to break down CCMP into its individual parts to simplify and explain it. Recall that *counter mode* (CM) is one of the modes in which block ciphers can operate. You might also recall the *cipher-block chaining* (CBC) mode, another popular mode for block ciphers. This leaves us with the *Message Authentication Code* component of the name. We haven't talked about this yet so I'll briefly introduce it and its associated concepts next.

Hash Functions

Hash functions, or algorithms, are used for detecting unauthorized modifications to data. They serve as a type of watermark or digital signature that can be applied to data.

A popular example of a hash function is a one-way hash, in which a sending entity takes any arbitrary data, runs it through the hash function, and produces a fixed-length value called the *hash*. The resulting hash value is appended to the data and sent to the receiving entity. The receiver gets the data with the appended hash, runs the same hashing function on the data component, and compares the result with the hash value that was sent with the original data. If the two values are different, the receiver can conclude that the data may have been altered along the way. But if the two values are the same, the receiver can reasonably conclude that the integrity of the data has been preserved. The one-way hash function does not use any keys and is not used for encryption in any way.

NOTE The process of applying the hash function is sometimes called *digesting*, and the result is sometimes called a *message digest.*

Message authentication is a method used in cryptosystems for verifying the authenticity and integrity of data. The integrity aspects of message authentication are concerned with making sure that data is not modified or altered in any way before reaching its intended recipient. And the authenticity aspect is concerned with making sure that the data originates from the entity that receiver is expecting it to originate from. It is also referred to as *message integrity code* and *message authentication code*. Currently two approaches are used to ensure integrity and authenticity: Hash-based Message Authentication Code (HMAC) and Cipher Block Chaining Message Authentication Code (CBC-MAC).

HMAC   HMAC combines the process of hash functions with a Message Authentication Code function of some sort. One key difference between plain hash functions and the HMAC function is the use of a secret key. The key type used here is symmetric. The HMAC process works like this:

1. The sender and the receiver entities agree on a secret key to be used.

2. The sender decides to transmit some arbitrary data to a receiver. But the sender wants to make sure that data is received intact at the other end and wants to provide some guarantee to the receiver that the sender actually sent the data.

3. The sender appends the agreed upon secret key to the data to be sent.

4. The result is passed through a hash function. The new hash value is the message authentication code.

5. The message authentication code is appended to the original data (plain-text) and sent to the receiver.

6. The next steps are carried out at the receiver end.

7. The receiver receives the data with the appended message authentication code.

8. The receiver appends the shared secret key to the data. The result is run through a hash function again. The hash value is the receiver's version of the message authentication code.

9. The receiver compares his/her value of the message authentication code with the value that was received with the message from the sender.

10. If the computed value is the same as the received value, the data can be assumed to have passed the integrity and authentication test.

11. If the computed value is different from the received value, it can be assumed that the data was tampered with along the way.

Note that nowhere in the process is the data encrypted. Encryption or confidentiality is not a function of HMAC.

CBC-MAC   CBC-MAC is a combination of the CBC mode used in block ciphers and a message authentication code of some sort.

CBC-MAC works in a simple but elegant way, described as follows:

1. The sender and the receiver entities agree on a secret key to be used.

2. The sender decides to transmit some arbitrary data to a receiver. But the sender wants to make sure that data is received intact at the other end and also wants to provide some guarantee to the receiver that the sender actually sent the data.

3. The sender encrypts the arbitrary plain-text data using the CBC symmetric block cipher mode of operation. The encryption will be done on blocks of the plain-text and will result in corresponding blocks of cipher-text.

4. The output of the final block of cipher-text is used as the message authentication code, which is appended to the plain-text data.

5. The sender sends the plain-text and message authentication code combination to the receiver.

6. The next steps are carried out at the receiver end.

7. The receiver encrypts the plain-text data with the shared secret key. The encryption is performed with the CBC symmetric block cipher again, resulting in blocks of cipher-text.

8. The last block of cipher-text is used as the receiver-generated message authentication code value.

9. The receiver compares the message authentication code value with the value that was received with the message from the sender.

10. If the computed value is the same as the received value, the data can be assumed to have passed the integrity and authentication test.

11. If the computed value is different from the received value, it can be assumed that the data was tampered with along the way.

Note that nowhere in the process is the hashing algorithm performed. This is the key distinguishing factor between HMAC and CBC-MAC. You should also note that the whole point of the CBC-MAC process is not to hide the plain-text data.

EAP

The Extensible Authentication Protocol (EAP) is an authentication framework that forms a basis upon which other authentication schemes rest. It forms a big part of the ratified IEEE 802.11i standard.

As indicated by its name, the EAP framework was designed with extensibility and authentication in mind. For the authentication component, various credentials such as usernames, passphrases, digital certificates, smart cards, and one-time passwords are acceptable.

The extensibility component of EAP implies several things as follows:

■ Arbitrary authentication mechanisms can be designed around it.

■ The base EAP specification can be kept simple and lightweight.

■ Advanced features and their resulting complexity can be abstracted away from EAP and implemented in the solutions that use EAP.

■ EAP is relatively future-proof. As technology advances and new demands are placed on existing solutions, EAP can easily adapt to accommodate these changes.

EAP can be encapsulated inside any Data Link layer protocol such as Ethernet, Point-Point-Protocol (PPP), IEEE 802.11 frames, and so on.

EAP Entities

Entities are the network components that use EAP to meet their authentication and key management needs in a wireless network. Notice the close semblance and verbiage to the components in the RADIUS world (see Chapter 10) or the IEEE 802.1X world. The entities in any EAP scheme may include any of the following components:

- **Peer**   The device or the user that wants to access the protected network resources. In a wireless network environment, the peer is often the wireless STA, the entity that responds to the authenticator. In the IEEE-802.1X world, this component is also known as the supplicant.

- **Authenticator**   The gatekeeper entity that initiates the EAP authentication conversation with the back-end components. In a wireless network environment, this could be a wireless access point (WAP). In the RADIUS world, it is called the network access server (NAS).

- **Backend authentication server**   The authenticator relies on the back-end authentication server to provide authentication services. This component executes EAP methods or grammar on behalf of the authenticator.

- **EAP server**   The component that terminates the EAP conversation with the peer component. This component or functionality is often discretely packaged with the back-end authentication component. When combined in this way, the service they provide is similar to the service provided by the access server in the RADIUS world.

The back-end authentication server and the EAP server possess the final knowledge of who should have access to what and when.

EAP Grammar

EAP *grammar* refers to the way EAP is spoken among the components that want to use the authentication and key management services provided by EAP. EAP's grammar is quite simple and consists of the following four primitives:

- **Request**   The authenticator sends these types of packets to the supplicant.
- **Response**   The supplicant sends these types of packets to the authenticator.
- **Success**   These are used to indicate successful authentication.
- **Failure**   These are used to indicate an unsuccessful authentication.

EAP Types

Different implementations of the EAP framework exist to address the issues of authentication and key exchange. Note that the EAP framework is concerned not only with securing wireless communications but also with wired communications. The following sections discuss some EAP implementations that are especially popular in the wireless world.

EAP-TLS

The EAP-Transport Layer Security (EAP-TLS) authentication protocol was developed by Microsoft and is fully described in the IETF's RFC 5216. It uses the facilities provided in the traditional standalone TLS protocol.

EAP-TLS relies on the use of digital certificates to authenticate the parties that want to communicate with one another. It requires that the entities that want to use EAP for authentication and key management mutually authenticate each other. This means that the entities participating in the EAP-TLS conversation require their own digital certificates. In environments where digital certificates are not already widely in use, this requirement may add an extra burden for the wireless network administrator.

EAP-TLS Conversation   A successful EAP-TLS conversation between EAP entities is outlined in the following steps. The entities are the peer (such as a wireless client STA), the authenticator (such as a WAP), and the EAP server/back-end authentication server (such as a RADIUS server).

1. The parties that want to be authenticated—the peer and the authenticator— agree to do so via EAP.

2. The authenticator asks the peer to identify itself via an EAP-Request message.

3. The peer sends its identity to the authenticator via an EAP-Response message.

4. From this point on, the EAP-TLS conversation appears to happen between the peer and the authenticator. But in reality, the authenticator component is acting as a go-between, by conveying the messages between the peer and the EAP server/back-end authentication server components.

5. After receiving the peer's identity, the EAP server sends (via the authenticator) an EAP-Request message with the EAP-Type set to EAP-TLS. This message serves as a way of telling the peer that the EAP-TLS conversation is about to begin.

6. The peer responds with an EAP-Response packet with the EAP-Type set to EAP-TLS. The packet will serve as a way for the peer to agree to start EAP-TLS via a client_hello message.

7. The EAP server responds to the peer with an EAP-Request packet with the EAP-Type set to EAP-TLS. The packet contains information such as the server TLS certificate and the server_hello message. The server will also request the peer to send its own certificate.

8. The peer responds with an EAP-Response packet with the EAP-Type set to EAP-TLS. The packet contains information such as the peer's TLS certificate and so on.

9. The EAP server verifies the peer's certificate and digital signature. The peer does the same thing.

10. If everything checks out from the server's perspective, the EAP server sends an EAP-Success packet.

11. If the peer fails to authenticate itself successfully to the EAP server, the server sends an EAP-Failure message.

EAP-TTLS

The EAP-Tunneled Transport Layer Security (EAP-TTLS) protocol is an extension of the EAP-TLS mechanism that is described in the IETF's RFC 5281.

EAP-TTLS is different from EAP-TLS because it does away with the EAP-TLS requirement of a supplicant-side certificate. Only the authentication server component requires a digital certificate.

The authentication server is authenticated using its digital certificate. An encrypted tunnel is then established between the peer (or supplicant) and the authentication server. The peer's authentication credentials, such as a digital certificate or password, are passed to the authentication server over the established tunnel. The peer can use other authentication methods such as Challenge-Handshake Authentication Protocol (CHAP), Password Authentication Protocol (PAP), and Microsoft CHAP (MS-CHAP) v2. These alternatives were discussed in Chapter 10.

Having to manage certificates only on the server side makes EAP-TTLS much easier to manage, because the wireless administrator does not have to worry about creating and managing digital certificates on all the wireless client STAs.

EAP-PSK

The EAP Pre-Shared Key authentication protocol, like the other EAP types discussed thus far, can be used for providing authentication services to entities in a wireless network. It is described in the IETF document RFC 4764.

The Pre-Shared Key refers to a key or secret that needs to be derived and shared by the parties by some mechanism before the EAP-PSK conversation takes place. The security provided by EAP-PSK will be compromised if this secret key is exposed.

Note that EAP-PSK is different from the Pre-shared Key authentication mode used in Wi-Fi Protected Access (WPA). The WPA implementation is commonly known as WPA-PSK.

EAP-PSK is a simpler in its design and in the way it functions when compared to EAP-TLS or EAP-TTLS. This simplicity is due to the fact that it does not use asymmetric cryptography as the other two do. This same simplicity also means that EAP-PSK cannot offer some of the advanced security features of the others.

Under the hood, EAP-PSK uses the AES symmetric block cipher.

EAP-SIM

This EAP authentication type uses the Subscriber Identity Module (SIM) used on Global System for Mobile (GSM) communications mobile networks. GSM is a second-generation mobile network standard. (See Chapter 7 for more on GSM.) The SIM card, as it is fondly called, is a type of smart card that is distributed by the mobile network operators to their subscribers.

EAP-SIM was developed by the Third Generation Partnership Project (3GPP). The IETF document RFC 4186 describes EAP-SIM.

EAP-SIM builds on and extends the traditional GSM security mechanisms. It extends normal GSM authentication by providing a mechanism for the parties to authenticate each other mutually.

EAP-AKA

The EAP-Authentication and Key Agreement (AKA) protocol is used in third-generation (3G) mobile networks. Universal Mobile Telecommunications System (UMTS) and CDMA2000 are regarded as 3G mobile technology standards.

EAP-AKA can be used for authentication purposes along with 3G identity modules and network infrastructures.

EAP-AKA was developed by 3GPP. The IETF document RFC 4187 describes EAP-AKA.

In general, EAP-AKA provides more advanced security mechanism compared to EAP-SIM.

NOTE   It is possible to use EAS-SIM or EAS-AKA to authenticate to noncellular networks such as wireless local area networks (WLANs).

All that is needed is to configure the authenticator (WAP) to send authentication requests to back-end authentication servers or the EAP server of a cellular network provider.

For example, the owner (the supplicant) of a GSM phone could use her credentials, which are stored on her SIM card, to hop onto a foreign WLAN. As long as the WLAN is preconfigured for this, the cellular network provider will bill the user on behalf of the WLAN operator for the use of the high-speed WLAN network. The user gets the benefits of using the faster WLAN resources for her data transfers (or even voice over IP calls) instead of using the more expensive provider cellular network. The role of the cellular service provider in this case is relegated to mere authentication and accounting functions.

IEEE 802.11i

This chapter has been building toward this discussion—almost everything discussed so far will help you understand the why and the how behind IEEE 802.11i.

The why was answered in the discussion about WEP and WPA. WEP was one of the earlier attempts at providing some kind of security for wireless communications.

In due time, several vulnerabilities were discovered in WEP that made it no longer suitable. However, WEP was so widely engrained in so many wireless network security solutions that great care had to be taken to provide an alternative for it. This was especially important because proposed alternatives had to work with existing and widely deployed wireless hardware. The stop-gap solution was WPA, and the requirements for WPA were designed such that existing equipment that use WEP could be easily upgraded to support WPA via software or firmware updates. Eventually, WPA2 was finalized and was a big part of the IEEE 802.11i picture.

The how behind IEEE 802.11i is in all the pieces that work together to offer a long-term authentication, confidentiality, and integrity as a security solution for wireless networks. Some of the pieces are 802.1X/EAP, which is used for authentication, and the AES-CCMP, which is used for satisfying the integrity and confidentiality needs.

Four-Way Handshake

An important aspect of the different EAP types was to provide a secure means of authenticating the parties that wanted to communicate. Another important by-product of the EAP is to generate symmetric keys, such as the Master Session Key (MSK)—an all-important key.

After the authentication stage has been successfully completed, next comes the four-way handshake, an Authentication and Key Management Protocol (AKMP) used in IEEE 802.11i. Its job is to confirm that the parties that want to communicate securely each possess the Pairwise Master Key (PMK) and to also distribute the group keys. The PMK is derived from the MSK.

In general, the so-called master keys are not themselves used for encrypting data. They are used for generating other subordinate and temporary keys that can be used for encrypting data.

The four-way handshake is used for generating dynamic keys that will be used for protecting subsequent data transmissions. These keys are transient or temporary by nature and as such are referred to as *transient keys* and *temporal keys*. The two types of transient keys that can be derived from the four-way handshake are the Pairwise Transient Key (PTK) and the Group Temporal Key (GTK).

In general, the pairwise keys are used only between a pair of communicating entities. The group keys can be used between two or more communicating entities.

IEEE 802.11i Considerations

With all the terms and concepts used in this chapter, you may not be immediately clear about what exactly you, as a wireless network administrator, need to do to deploy a network that is IEEE 802.11i compliant. Table 11-2 presents that information in bite-sized pieces.

Scenario	Suggested Solutions and Tips
You manage a small wireless network with only a few users and devices and need to use the most current security mechanisms.	Consider IEEE 802.11i. Use IEEE 802.11i with pre-shared keys, which help you avoid the additional overhead of maintaining a public key infrastructure (PKI).
You manage a large to medium-sized wireless network and need to use the most current security mechanisms.	Consider IEEE 802.11i. Consider using IEEE 802.11i with one of the EAP authentication types.
You already use digital certificates for server- or infrastructure-side authentication on your network.	Consider using EAP-TTLS with IEEE 802.11i. EAP-TTLS does *not* require mutual authentication.
You already use digital certificates for server- or infrastructure-side authentication and for client-side authentication on your network.	Consider using EAP-TLS with IEEE 802.11i. EAP-TLS requires mutual authentication.
You are considering using EAP-TLS or EAP-TTLS on the infrastructure side of your wireless network.	You need a RADIUS server implementation that supports EAP. You need a PKI that lets you manage the digital certificates that will be used, which includes the tasks of issuing, revoking, signing, and storing the certificates. Some of this can occur via a certificate authority implementation. Your WAPs (the authenticator entity) must support 802.1X so that they can communicate with the RADIUS servers via EAP. You might get 802.1X support in infrastructure hardware via simple firmware upgrades if the vendor doesn't support it. Consider temporarily allowing legacy authentication mechanism to coexist with new 802.1X-based mechanisms, which will help if things don't work as smoothly as planned.
You are considering using EAP-TLS or EAP-TTLS on the client or peer side of the wireless network.	You need the wireless clients STAs to be able to talk 802.1X to the access point. The supplicant software on the clients needs to support the EAP type with which the server is configured.

Table 11-2.   IEEE 802.11i Wireless Network Deployment Considerations

Summary

This was a tough chapter to write and probably a tough read, too. Weaving all the sometimes disparate topics together to try to build one simple idea is a difficult task.

Cryptography is not for the faint of heart; nor is it an easy subject. Its beauty lies in the fact that it is guided by simple core principles, which have remained unchanged for centuries, and they are mostly common sense principles.

You learned that any good cryptosystem should provide authentication, confidentiality, and integrity.

The methods (or algorithms) that have been developed to provide or support these objectives were discussed. We started by looking at some generic algorithm/ciphers that form the bedrock of modern-day cryptographic systems.

We moved on to practical applications of the low-level concepts and algorithms that have been reused in developing standards and protocols for protecting wireless networks. We glued all the theoretical aspects together with the practical aspects of administering a secure wireless network.

After all has been said and done, remember that in matters of secrecy, *"Three may keep a secret, if two of them are dead."*

PART V | Wireless Devices Configuration and Other Wireless Network Considerations

CHAPTER 12 | Infrastructure Device Configuration

Key Skills and Concepts

- Understand the key infrastructure components of a wireless network.
- Understand the functions and roles of components of a wireless network.
- Design a simple wireless local area network (WLAN).
- Configure a generic wireless access point (WAP) device.
- Configure a generic WLAN controller device.

I n this chapter, we begin building a simple wireless network that will incorporate some of the elements and technologies discussed in the preceding chapters.

Our sample network will comprise infrastructure-side devices and client-side devices. The infrastructure component of our network will include the following:

- A standalone WAP
- A WLAN controller
- Managed WAPs

And the following devices make up the client components:

- Windows-based client devices
- Macintosh-based client devices
- Linux-based client devices

We will take a four-step approach to building this heterogeneous network:

1. Set up the infrastructure side devices in this chapter.
2. Set up the Windows-based client device in Chapter 13.
3. Set up the Macintosh-based client device in Chapter 14.
4. Set up the Linux-based client device in Chapter 15.

We will build our network to conform to the network diagram shown in Figure 12-1. We'll begin building our infrastructure network by identifying the types and roles of the individual devices that make up the network. Then we'll walk through configuring the devices.

NOTE   Throughout this part of the book, the configuration process has been kept as generic as possible so that you can use the knowledge gained here to configure other real-world infrastructure devices. The basic principles and concepts guiding the functioning and configuration of wireless networking gear are similar; what differs sometimes is the terminology and user interfaces that the equipment manufacturers use in their products.

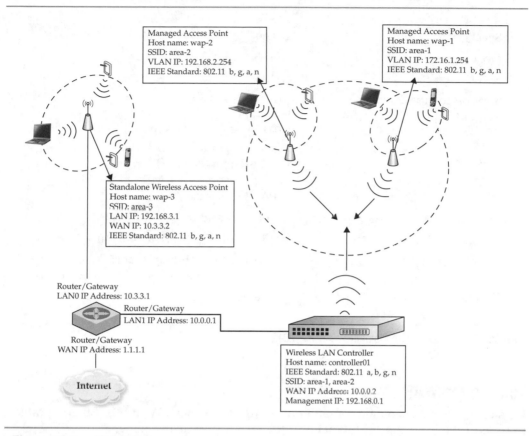

Figure 12-1. Network diagram

The Wireless Access Point

The network WAP in our sample network functions in *standalone mode,* which means that it is not being managed by a central controller. It does not take its configuration information from any other device; all configuration is performed directly on the device by the wireless network administrator.

The WAP in our network is the award-winning model nix-WAP-2012 (a fictitious product) made by Caffenix, with the following outstanding features:

- Rugged and works under extreme conditions.

- Easy to configure.

- Standards-based.

- It's free.

The specifications of the nix-WAP-2012 and our configuration objectives are shown in Table 12-1.

Specification	Value
Host name	wap-3
SSID	area-3
Supported ciphers and authentication	WEP 64 / 128 bits WPA-Personal / TKIP WPA-Personal / CCMP WPA-Enterprise / TKIP WPA2-Personal / TKIP WPA2-Personal / CCMP WPA2-Enterprise / CCM
Encryption key or passphrase	never-never-land
Wireless PHYs supported	802.11b, 802.11g, 802.11a, 802.11n
Physical ports	1 10/100 Mbit/s Ethernet port (WAN port) 4 Gigabit Ethernet ports (LAN switch ports)
Wireless interface	1 wireless interface

Table 12-1.   Specifications of the nix-WAP-2012

Configuring the WAP

We start by using a separate computer to configure the WAP. We use an Ethernet cable to connect the computer's Ethernet port to any of the local area network (LAN) ports of the WAP. By default, the WAP automatically assigns an IP address to any system plugged into any of its LAN ports.

We log into the management interface of the WAP using a web browser. Out of the box, the WAP has a factory-assigned IP address of 192.168.3.1, and it also has a factory-assigned administrative username of "administrator" and a very *insecure* password of "password."

After we log in, the nix-WAP-2012 prompts us to launch the initial configuration wizard. The entire configuration process for the WAP is separated into three areas. The wizard also separates the configuration tasks along these three categories:

- Global Configuration
- Network Configuration
- Wireless and Security Configuration

The wizard uses a question-and-answer approach to make the initial configuration of the unit as easy as possible. It asks questions and uses your answers to set the configuration parameters. Some of the questions are helpfully interspersed with explanations.

The following sections walk through the process and options selected while we use the wizard.

Global Configuration of the nix-WAP-2012

The wizard starts with the global configuration tasks.

1. Select the option to start the configuration wizard for the AP.

2. Create a host name or system name for the WAP. Type the host name shown here and then press ENTER.

   ```
   > wap-3
   ```

3. Change the default factory-assigned administrative password for the WAP. Type this new password:

   ```
   > caffenix-wap3-pw
   ```

4. Would you like to save the changes to the global configuration section?

   The options are {YES} {NO}

   ```
   > YES
   ```

CAUTION   It is very important that you always change the passwords for all networking equipment from the default factory-assigned values when initially configuring such devices. Not doing so can have serious security consequences.

Network Configuration of the nix-WAP-2012

The network configuration section of the wizard setup is where Internet Protocol (IP) addresses are assigned to the ports and interfaces on the unit. The gateway, Dynamic Host Configuration Protocol (DHCP), and Domain Name System (DNS) values are also set.

Figure 12-2 shows an exploded view of the interfaces available on the nix-WAP-2012 and the network addresses that will be assigned to the interfaces when we are done.

The wizard continues to the Network Configuration tasks:

1. How will the WAN port be configured?

   The options are {static IP} {Automatic – DHCP} {PPPOE}

   ```
   > static IP
   ```

2. What is the IP address and network mask for the WAN interface? Press ENTER when done.

   Hint: The accepted format is xxx.xxx.xxx.xxx / xxx.xxx.xxx.xxx

   ```
   > 10.3.3.2 /  255.255.255.252
   ```

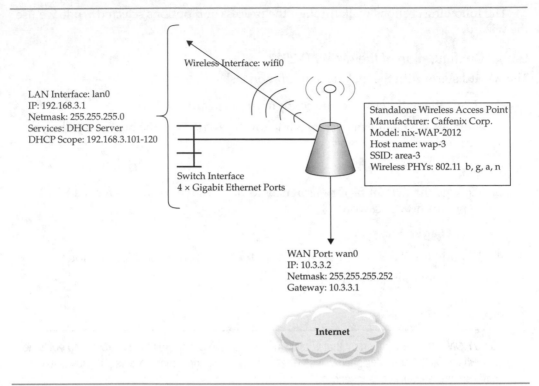

LAN Interface: lan0
IP: 192.168.3.1
Netmask: 255.255.255.0
Services: DHCP Server
DHCP Scope: 192.168.3.101-120

Wireless Interface: wifi0

Standalone Wireless Access Point
Manufacturer: Caffenix Corp.
Model: nix-WAP-2012
Host name: wap-3
SSID: area-3
Wireless PHYs: 802.11 b, g, a, n

Switch Interface
4 × Gigabit Ethernet Ports

WAN Port: wan0
IP: 10.3.3.2
Netmask: 255.255.255.252
Gateway: 10.3.3.1

Internet

Figure 12-2.   Exploded view of interfaces and network addresses

3. Enter the default remote gateway (or router) IP address. Press ENTER when done.

   Hint: The accepted format is xxx.xxx.xxx.xxx / xxx.xxx.xxx.xxx

   ```
   > 10.3.3.1 / 255.255.255.252
   ```

4. Enter the remote DNS server IP address for this WAP. Press ENTER when done.

   Hint: The accepted format is xxx.xxx.xxx.xxx / xxx.xxx.xxx.xxx

   ```
   > 10.3.3.1 / 255.255.255.252
   ```

5. Enter the IP address and network mask for the LAN interface. Press ENTER when done.

   Hints: The wireless interface (wifi0) is bridged or wired internally to the four LAN switch ports to form a single virtual interface. This new virtual interface is called the LAN interface of the access point (AP) and is treated as such. This combined interface name is lan0.
   The accepted format is xxx.xxx.xxx.xxx / xxx.xxx.xxx.xxx

   ```
   > 192.168.3.1 /  255.255.255.0
   ```

6. Do you want this device to automatically assign IP addresses to clients connected to its wireless and Ethernet interfaces?

    The options are {YES} {NO}

    Hint: Answering YES here will turn on the DHCP server service on the AP. To avoid conflicts on any other directly connected networks, make sure that only one DHCP server is running on any network segment.

    ```
    > YES
    ```

7. By default, the DHCP server you have just enabled can assign 255 unique IP addresses to DHCP clients. You can change this default value by supplying a lower range. Would you like to change the default DHCP range?

    The options are {YES} {NO}

    Hint: This DHCP range is also known as the DHCP scope.

    ```
    > YES
    ```

8. Enter the new DHCP starting address range. Press ENTER when done.

    Hints: The starting range is the lower boundary of the DHCP scope or the smallest IP address in the scope.
    The accepted format is xxx.xxx.xxx.xxx

    ```
    > 192.168.3.101
    ```

9. Enter the new DHCP ending address range. Press ENTER when done.

    Hints: The ending range is the upper boundary of the DHCP scope Or the highest IP address in the scope.
    The accepted format is xxx.xxx.xxx.xxx

    ```
    > 192.168.3.120
    ```

10. Your DHCP server will be able to supply 20 DHCP clients at a time with IP configuration information. Is this OK? Press ENTER when done.

    The options are {YES} {NO}

    ```
    > YES
    ```

11. Would you like to save the changes to the Network Configuration section?

    The options are {YES} {NO}

    ```
    > YES
    ```

Wireless and Security Configuration of the nix-WAP-2012

The wireless and security configuration section of the WAP wizard setup controls the radio frequency (RF) properties of the unit. The WAP itself should, of course, have the built-in hardware to support the different PHY technologies that we will be configuring.

The wizard continues to the Wireless Configuration tasks:

1. This WAP has built-in wireless radios that can support the following wireless technologies: 802.11b, g, a, n. Which of the wireless technologies would you like to enable? Press ENTER when done.

   The options are {802.11b} {802.11g} {802.11a} {802.11n}

   Hint: You should enable support only for the technologies that you use.

   > 802.11n

2. Create a network name (SSID) for the wireless network.

   > area-3

3. Would you like to secure communication for the wireless network?

   The options are {YES} {NO}

   > YES

4. There are several authentication and encryption methods that can be used to secure the wireless network. Which of these would you like to use?

   The options are {WEP 64 bits} {WEP 128 Bits} {WPA-Personal/TKIP}

   Hint: Make sure that whatever option you select is also supported by the wireless clients.

   > WPA-Personal/TKIP

5. Please create a secret key that will be used in encrypting and decrypting the communications. Press ENTER when done.

   Hint: The secret key is the passphrase. A good passphrase should be a random alphanumeric string at least eight characters long.

   > never-never-land

6. Would you like to save the changes to the network configuration section?

   The options are {YES} {NO}

   > YES

The WLAN Controller

The second major device in our infrastructure arsenal is the WLAN controller, which can be used to manage various aspects of the APs that it controls. For example, the controller can be used to perform configuration, firmware, radio resource management, auditing, and security functions of the APs connected to it.

Our WLAN controller of choice is the award-winning model nix-WLC-2012 made by Caffenix. The nix-WLC-2012 has won many industry awards and is the best in its class. Following are the highlights of the unit:

- Rugged and works under extreme conditions
- Modular
- Easy to configure
- Standards-based
- It's free

The capabilities and specifications of the nix-WLC-2012 are shown in Table 12-2, which also shows our configuration objectives.

Configuring the WLAN Controller

We will use a separate computer to configure the controller. We will connect the Ethernet port of the computer to the management port of the controller using an Ethernet cable. The controller will automatically assign an IP address to any system connected to its management port.

We log into the management interface of the WLAN controller using a web browser. Out of the box, the management interface has a factory-assigned IP address of 192.168.0.1. The controller also has a factory-assigned administrative username of "administrator" and a very *insecure* password of "password."

Specification	Value
Host name	controller01
Supported ciphers and authentication	WEP 64 / 128 bits WPA-Personal / TKIP WPA-Personal / CCMP WPA-Enterprise / TKIP WPA2-Personal / TKIP WPA2-Personal / CCMP WPA2-Enterprise / CCMP
Physical ports	1 Gigabit Ethernet 1 management port
Virtual/logical interfaces supported	1–100
Wireless PHYs supported	802.11b, 802.11g, 802.11a, 802.11n, Bluetooth, ZigBee, RFID, GSM, CDMA

Table 12-2. WLAN Controller Specifications

After we log in, the controller will prompt us to launch the initial configuration wizard. The entire configuration process for the controller is separated into five areas. The wizard also separates the configuration tasks along these five areas:

- Global Configuration
- Network Configuration
- Wireless Configuration
- Virtual Local Area Network (VLAN) Configuration
- VLAN Security Configuration

The wizard uses a question-and-answer approach to make the initial configuration of the unit as easy as possible. It asks the questions and uses the answers supplied by the user to set the configuration parameters. Some of the questions are helpfully interspersed with explanations called *hints*.

Global Configuration for nix-WLC-2012

The following steps walk through the process and options selected while using the wizard. The wizard starts off with the global configuration tasks.

1. Select the option to start the configuration wizard for the controller.

2. We need to create a host name or system name for the controller. Type the host name shown here when prompted by the wizard:

   ```
   > controller01
   ```

3. Create a user account that will be used to manage the controller. Type this username when prompted:

   ```
   > admin
   ```

4. We need to change the default factory-assigned administrative password for the controller. Changing the controller's administrative password will also automatically update the administrative passwords of all the devices that are managed by this controller. Type this password when prompted by the wizard:

   ```
   > caffenix-controller-pw
   ```

5. Would you like to save the changes to the global configuration section?

   The options are {YES} {NO}

   ```
   > YES
   ```

Network Configuration for the nix-WLC-2012

The Network Configuration section of the setup wizard deals with the networking properties of physical interfaces or ports on the nix-WLC-2012. This is where IP addresses are assigned to the ports and interfaces on the unit. This is also where the gateway, DHCP, and DNS values are set.

The settings here affect how the controller connects to the wired network segment, and this, by extension, affects how the remotely managed APs reach the wired segment, too.

Figure 12-3 shows an exploded view of the physical and logical interfaces available on the nix-WLC-2012 and the network addresses that will be assigned to the interfaces when we are done.

The wizard continues to the Network Configuration tasks:

1. How will the WAN port be configured?

    Hint: The WAN port is used for connecting the controller to other wired networks.

    The options are {static IP} {Automatic – DHCP}

    ```
    > static IP
    ```

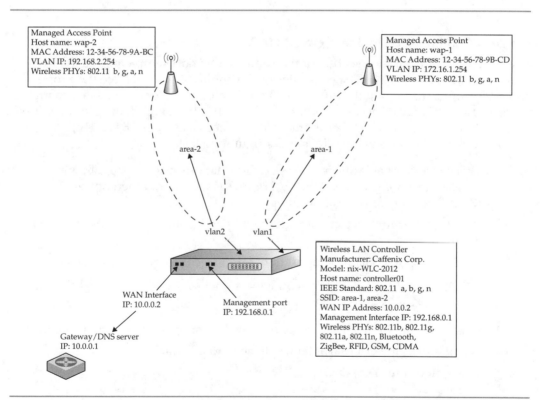

Figure 12-3.   Exploded view of the physical and logical interfaces available on the nix-WLC-2012

2. What is the IP address and network mask for the WAN interface? Press ENTER when done.

   Hint: The accepted format is xxx.xxx.xxx.xxx / xxx.xxx.xxx.xxx

   ```
   > 10.0.0.2 / 255.255.255.252
   ```

3. Enter the default gateway IP address for the controller. Press ENTER when done.

   Hint: The accepted format is xxx.xxx.xxx.xxx / xxx.xxx.xxx.xxx

   ```
   > 10.0.0.1 / 255.255.255.0
   ```

4. Enter the DNS server IP address for the controller. Press ENTER when done.

   Hint: The accepted format is xxx.xxx.xxx.xxx / xxx.xxx.xxx.xxx

   ```
   > 10.0.0.1 / 255.255.255.0
   ```

5. Would you like to save the changes to the network configuration section?

   The options are {YES} {NO}

   ```
   > YES
   ```

Wireless Configuration for the nix-WLC-2012

The Wireless Configuration section of the controller wizard setup controls the RF properties of the controller. The controller itself should, of course, have the proper hardware built in to support the different PHY technologies. If, for example the controller needs to manage an AP that supports only the IEEE 802.11n standard, the controller itself needs to have a built-in radio that can transmit and receive using IEEE 802.11n.

The wizard continues to the Wireless Configuration tasks:

1. This controller has built-in wireless radios that can support the following wireless technologies: 802.11b, 802.11g, 802.11a, 802.11n, Bluetooth, ZigBee, RFID. These are also the allowed options.

   Which of the wireless technologies would you like to enable? Press ENTER when done.

   Hint: You should enable support only for the technologies that you use.

   The options are {802.11b} {802.11g} {802.11a} {802.11n} {blue} {zee} {rfid}

   ```
   >   802.11b   802.11g   802.11a   802.11n
   ```

2. This controller has internal antennas as well as support for external antennas.

   Will there be any external antenna attached to the unit?

   The options are {YES} {NO}

   ```
   > YES
   ```

3. Would you like to save the changes to the Wireless Configuration section?

   The options are {YES} {NO}

   ```
   > YES
   ```

Virtual Local Area Network (VLAN) Configuration for the nix-WLC-2012

A VLAN in our wireless controller terminology is like a virtual interface that is used for communicating with a particular WAP that is to be managed by the controller. Our sample wireless LAN controller can support up to 100 different VLANS.

Each VLAN is in its own little world, and the default behavior of the controller is to disable all communication between different VLANs on the controller. Permissions must be explicitly granted for any communication to take place between all the APs on a controller.

The virtual interface associated with each VLAN is named vlan*x*, where *x* is a number corresponding to the virtual interface. So, for example, for the first VLAN to be created, the interface will be named vlan1, and the second will be named vlan2, and so on.

The Media Access Control (MAC) address of the managed AP is used as a unique identifier when associating the AP with the VLAN defined on the controller.

With this background information we can now continue with the configuration of the controller.

1. Type the number of access points that this controller will manage now. Press ENTER when done.

   The options are {1} {2} {3} {4}...{100}

   > 2

2. Type the name of the VLANs to be associated with each access point. Press ENTER when done.

   The options are {vlan1} {vlan2} {vlan3} {vlan4}...{vlan100}

   > vlan1   vlan2

3. Enter the MAC address of the access point to be associated with vlan1. Press ENTER when done.

   > 12-34-56-78-9A-BC

4. Enter the MAC address of the access point to be associated with vlan2. Press ENTER when done.

   > 12-34-56-78-9B-CD

5. Would you like to configure vlan1 now? Press ENTER when done.

   The options are {YES} {NO}

   > YES

6. Enter the SSID to be associated with vlan1. Press ENTER when done.

   > area-1

7. A logical interface has to be created for the remote access point connected to vlan1. The interface name will be vlan1.1. Specify an IP address and subnet mask for the interface for the remote AP (vlan1.1). Press ENTER when done.

Hint: The accepted format is xxx.xxx.xxx.xxx / xxx.xxx.xxx.xxx

```
> 172.16.1.254 / 255.255.255.0
```

8. Are you ready to save and apply the configuration for vlan1? Answering YES will push the configuration settings to the managed wireless access point. Press ENTER when done.

   The options are {YES} {NO}

   ```
   > YES
   ```

9. Would you like to configure vlan2 now?

   The options are {YES} {NO}

   ```
   > YES
   ```

10. Enter the SSID to be associated with vlan2. Press ENTER when done.

    ```
    > area-2
    ```

11. A logical interface has to be created for the remote access point connected to vlan2. The interface name will be vlan2.1. Specify an IP address and subnet mask for the interface for the remote AP (vlan2.1).

    Hint: The accepted format is xxx.xxx.xxx.xxx / xxx.xxx.xxx.xxx

    ```
    > 192.168.2.254 / 255.255.255.0
    ```

13. Are you ready to save and apply the configuration for vlan2? Answering YES will push the configuration settings to the managed wireless access point.

    The options are {YES} {NO}

    ```
    > YES
    ```

Control and Provisioning of Wireless Access Point Protocol

Our fictitious wireless LAN controller uses an oversimplified mechanism to find and establish initial communications with the WAPs.

The Control And Provisioning of Wireless Access Points (CAPWAP) protocol is an example of a real-world method (protocol) that does the same thing and much more. It is based on an IETF specification described in RFC 5414.

CAPWAP is an interoperable protocol that enables an access controller to manage, control, and provision a collection of wireless STAs (such as APs). It simplifies the deployment and management of wireless networks.

CAPWAP uses a simple mechanism to perform its functions. The WAPs that need to be managed send a discovery "Request" message. The controller responds with a discovery "Response" message. After this initial set of messages is sent back and forth, the real management process begins.

VLAN Security Configuration of the nix-WLC-2012 The preceding configuration tasks dealt with setting up the VLANs that are tied to the individual managed WAPs. This final section of the controller setup wizard is used for setting up the security parameters, such as authentication, authorization, accounting, and so on, to be used on the VLANs.

Note that the parameters set here affect the wireless STAs (clients) that will connect to the managed WAPs. So the controller (and the wireless network administrator) must consider the security capabilities of the wireless STAs.

The VLAN security configuration begins here:

1. This controller currently has two VLANs defined. Which of these would you like to configure now? Press ENTER when done.

   The options are {vlan1} {vlan2}

   Hint: The SSID for vlan1 is area-1 and SSID for vlan2 is area-2.

   ```
   > vlan1
   ```

2. Several authentication and encryption methods can be used to secure the area-1 wireless network. Which of these would you like to use?

   The options are {WEP 64 bits} {WEP 128 Bits} {WPA-Personal / TKIP}

   {WPA-Personal / CCMP} {WPA-Enterprise / TKIP} {WPA2-Personal / TKIP}

   {WPA2-Personal / CCMP} {WPA2-Enterprise / CCMP}

   Hints: Make sure that whatever option you select is also supported by the wireless clients.
   Separate multiple options with commas, spaces, or semicolons.

   ```
   > WPA-Personal / TKIP
   ```

3. Please create a secret key that will be used in encrypting and decrypting the communications on area-1. Press ENTER when done.

   Hint: The secret key is the passphrase. A good passphrase should be a mix of alphanumeric characters and symbols and at least eight characters long.

   ```
   > $never-never-land1$
   ```

4. Are you ready to save the VLAN security configuration for vlan1? Answering YES will push the configuration settings to the managed wireless access point.

   Hint: Make sure that the remote wireless access point is powered. Also make sure that the AP is within reasonable physical distance to this controller to ensure radio connectivity.

   The options are {YES} {NO}

   ```
   > YES
   ```

5. The security parameters for vlan2 have not yet been set. Would you like to configure security for vlan2 now? Press ENTER when done.

   The options are {YES} {NO}

   ```
   > YES
   ```

6. Several authentication and encryption methods can be used to secure the area-2 wireless network. Which of these would you like to use?

   The options are {WEP 64 bits} {WEP 128 Bits} {WPA-Personal / TKIP} {WPA-Personal / CCMP} {WPA-Enterprise / TKIP} {WPA2-Personal / TKIP} {WPA2-Personal / CCMP} {WPA2-Enterprise / CCMP}

   Hints: Make sure that whatever option you select is also supported by the wireless clients.
   Separate multiple options with commas, spaces, or semicolons.

   ```
   > WPA2-Personal/CCMP; WPA2-Personal/TKIP; WPA-Personal/CCMP
   ```

7. Please create a secret key that will be used in encrypting and decrypting the communications for area-2. Press ENTER when done.

   ```
   > !never-never-land2!
   ```

8. Are you ready to save the VLAN security configuration for vlan2? Answering YES will push the configuration settings to the managed wireless access point.

   Hint: Make sure that the remote wireless access point is powered. Also make sure that the AP is within reasonable physical distance to this controller to ensure radio connectivity.

   The options are {YES} {NO}

   ```
   > YES
   ```

And that's all there is to it. We are done configuring the infrastructure devices for our wireless network.

Summary

We began setting the stage for our enterprise-grade wireless network in this chapter. We configured the infrastructure-side components, including installing a few devices that are commonly found in large to medium-sized networks. Our infrastructure hardware offered all the features, such as price, modularity, ruggedness, and so on, that would make both the wireless network administrator and the upper management folks happy.

The infrastructure devices that we set up in this chapter will be used by our network of heterogeneous clients (Windows, Linux, and Macintosh clients) in the next three chapters. So it is very important that we get everything almost perfect in this chapter.

The main players on our infrastructure side were a standalone WAP (nix-WAP-2012), which is similar in features to the off-the-shelf residential gateways/firewalls that are found in many homes and small to medium-sized networks. The configuration process for our sample WAP is similar to real-world devices that serve the same purpose.

The other major player in the infrastructure was a wireless LAN controller (nix-WLC-2012) that can be used to manage up to 100 APs, but we used it to manage just two APs in our sample network. The managed APs will each serve a wireless network with two different SSIDs. We followed a generic configuration process while configuring the nix-WLC-2012. Again the concepts discussed while configuring this unit will carry over when you are configuring real-world wireless LAN controllers.

CHAPTER 13 | Microsoft Windows Clients

Key Skills and Concepts

- Learn how to configure the wireless hardware components in Windows.
- Learn how to configure the software aspects for wireless networking in Windows.
- Learn how to select optimal and compatible wireless settings. Use the netsh utility for wireless network management.
- Explore SoftAP and Virtual WiFi features in Windows.
- Use wireless mesh networking in Windows.
- Learn about personal area networking in Windows.

Needless to say, a vast majority of desktop workstations in the world run a Microsoft Windows-based operating system (OS). In this chapter, we will look at setting up such systems as wireless client stations (STAs).

Windows Client Configuration

We will configure a computer running the Microsoft Windows 7 OS to connect to a wireless network. Our sample system will be configured to connect to a generic wireless access point (WAP) that was set up previously.

Table 13-1 shows the properties of the generic WAP, and Table 13-2 shows the properties of wireless client STA.

Specification	Value
Host name	wap-3
SSID	area-3
Supported ciphers and authentication	WEP 64 / 128 bits WPA-Personal / TKIP WPA-Personal / CCMP WPA-Enterprise / TKIP WPA2-Personal / TKIP WPA2-Personal / CCMP WPA2-Enterprise / CCM
Encryption key or passphrase	never-never-land
Wireless PHYs supported	802.11b, 802.11g, 802.11a, 802.11n

Table 13-1.   WAP Specifications

Specification	Value
Host name	Windows7-A
Operating system	Microsoft Windows 7 Enterprise Edition
Driver	caffenix80211.sys
Supported ciphers and authentication	WEP WPA-Personal / TKIP WPA-Personal / CCMP WPA-Enterprise / TKIP
Compatible/optimal/suitable cipher and authentication	WPA-Personal / TKIP
Supported IEEE 802.11 standards	802.11b, 802.11g
Compatible/optimal/suitable IEEE 802.11 standard	802.11g
Wireless network name (SSID)	area-3

Table 13-2.   Wireless Client Specifications

Setting Up the Hardware

The rest of this chapter assumes that the wireless adapter on the client system is already detected by Windows and that the appropriate driver is already loaded. This will be the case for most original equipment manufacturer (OEM) computers (with wireless capabilities) that ship with the OS already preloaded. In some cases, however, a third-party wireless adapter must be used with an existing Windows-based system. It is also possible that the factory-installed adapter and/or driver will go wonky during the lifetime of the computer and will need to be replaced or reinstalled. In these kinds of situations, you need to know how to set up wireless hardware, and that's the focus of the following discussion.

Naturally, before we get to the stage of fiddling with the drivers, we have to physically insert the card into the system. Based on the various wireless adapter form factors, we have our work cut out for us in this regard.

For adapters that can be plugged into an external port on the system, such as USB and PCMCIA adapters, the physical task is simple.

For adapters that need to interface with the system via some internal and not so easy to access port, such as PCIe, Mini PCI, and PCI, the task may be a little more involved.

The manufacturer for our sample adapter is again Caffenix, which makes the best wireless adapters in the world. The wireless adapter model we'll use is caffenix-80211.

Let's walk through the steps required to install the driver software for our generic wireless adapter.

1. Follow the hardware manufacturer's instructions on how to properly physically insert the wireless adapter into the computer.

2. After the card has been installed, power on the computer and log into the system if necessary.

3. Wait for Windows to detect the new hardware and automatically attempt to locate and install the driver. If Windows finds the correct driver, go on to step 6.

4. If Windows fails to find and install the appropriate device driver automatically from any of its standard locations (the built-in driver cache, removable drives, or Windows Update site), the driver will have to be manually installed.

5. Make sure that you have the driver software available, either previously downloaded from the vendor website or prepackaged by the vendor along with the hardware on a media (such as a CD).

6. Click the Windows Start menu.

7. Type **hdwwiz.cpl** in the Search Programs And Files text box. Windows will return a list of items that are related to your search term.

8. From the returned list, click hdwwiz.cpl. The Device Manager window will appear.

9. Look for any device with an exclamation symbol (!) next to a name that is related to WLAN. Once you find the device, right-click the device name and choose Update Driver Software. An Update Driver Software dialog box will appear.

10. Click the Browse My Computer For Driver Software option. A new screen will appear, where you can point Windows to the exact location of the driver software that you have from step 5.

11. On our sample system, the driver software is located on a vendor-supplied disc in the D: drive of the computer. Specifically, the driver is located at D:\caffenix_Inc\Drivers\Windows7\X64\.

12. After you pointed Windows to the proper location of the driver, click Next.

13. Windows may prompt you with a Windows Security dialog box (see the following illustration) warning you that it can't verify the publisher of the driver software. Click the option Install This Driver Software Anyway to continue.

If you don't get any such warning, you may skip this step.

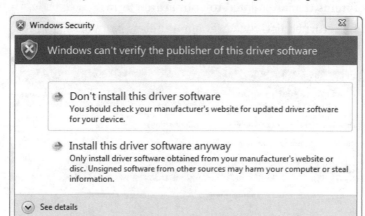

14. If all goes well, you should get a final screen informing you that "Windows has successfully updated your driver software." Click Close to finish.

15. The wireless adapter should now be ready for use.

Setting Up a Wireless Profile

After Windows has recognized the wireless adapter and has the proper drivers installed for communicating with the adapter, you can begin configuring the wireless client system running Windows by setting up a wireless profile.

You can connect a system to a wireless network in several ways in Windows 7. Windows can do this automatically, or you can do it manually. Allowing Windows to do the work can be easy and is a somewhat "dumbed-down" approach. In certain scenarios, it can also lead to unintended consequences (such as connecting to an unknown or hostile wireless network). We will opt for the manual approach here, because it allows us to have better control of the process and offers us some insight into what Windows is doing in the background.

The settings for individual wireless connections are stored in profiles in Windows 7. So if, for example, we want to set up a wireless connection to a wireless gateway at work, we could set up a profile specifically for this and call it "work." In addition to enabling connectivity with a wireless network, you can do many other cool things with profiles.

Let's walk through the process of creating a wireless connection profile on our sample system and then manually connect to the wireless network associated with the profile.

1. Log into the system with a user account with administrative privileges.

2. Open the Windows Start menu.

3. Type **wireless** in the Search Programs And Files text box. Windows will return a list of items that are related to your search term.

4. From the returned list, click Manage Wireless Networks. You'll see a window similar to the following:

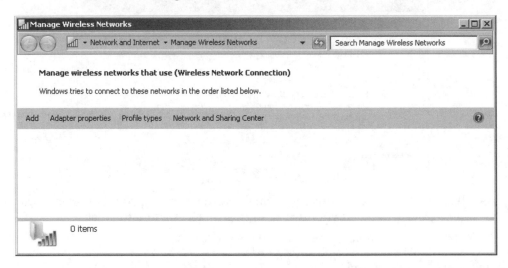

5. Click the Add button, and the Manually Connect To A Wireless Network dialog box will appear.

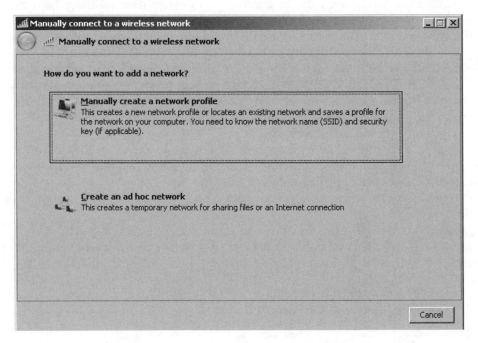

6. Click the Manually Create A Network Profile option. A screen similar to the one shown here will appear, allowing you to input the wireless settings.

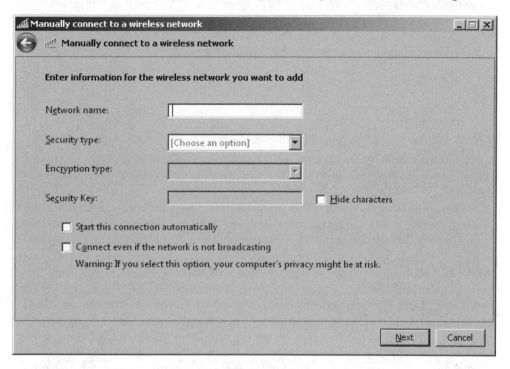

7. Complete the fields of the dialog box with the appropriate information provided in Table 13-2 earlier in the chapter.

8. For example, type **area-3** in the Network Name field (SSID). Use only the compatible/optimal/suitable values where appropriate. (See the sidebar titled "Compatibility, Optimization, and Suitability Considerations" for guidelines.)

9. Make sure that the check boxes beside these options are cleared: Start This Connection Automatically and Connect Even If The Network Is Not Broadcasting.

TIP Before the Windows client can successfully connect to a wireless infrastructure network where the AP does not advertise the SSID, the Connect Even If The Network Is Not Broadcasting option must be enabled, or checked.

10. Click Next when you have finished adding information to the fields.

11. At the ensuring Successfully Added "area-3" window, click Close.

NOTE   By default, Windows gives the profile created through this process the same name as the value of the SSID of the WLAN. If we wanted a more intuitive name, we'd have to rename the profile at the end of the profile creation process.

This could be done in the Manage Wireless Networks control panel, which is accessed by right-clicking the name of the network profile and selecting the Rename option.

Wireless Network Management Utilities: Windows vs. Third-party Software

We've set up a wireless profile using the built-in wireless network management utility in Windows 7. We can use other methods to achieve the same goal, however.

One such method is to use the management software provided by the wireless device's vendor or manufacturer to configure a wireless network interface on a Windows computer; Microsoft considers this method a "third-party solution."

When you install a vendor's wireless network management software, you might notice that you must choose to either use Windows' built-in wireless network management utility or the vendor-supplied wireless network management utility. You should either use one or the other, but not both. Windows will sometimes warn you if third-party software is currently being used for managing the wireless interface on a Windows system.

Wireless hardware vendors probably know the ins and outs of their hardware better than anybody else, including OS vendors. On the flip-side, OS vendors probably know better than anybody else how third-party hardware *should* interact with their OS internals. Herein lies the fork in the road.

Who should you trust to put in charge of managing the wireless network interfaces on your systems? Go with whichever one works and whichever one provides any extra functionality that you require.

If you want simplicity and seamlessness, use the built-in Windows utility. If you want greater control and extra bells and whistles that the vendor provides, stick with the vendor-supplied wireless network management software.

Despite the previous advice, you may sometimes find that in the real world, you don't have a choice between using a Windows built-in utility or a third-party utility, because only one of the solutions actually works. The built-in Windows utility may fail completely while the third-party utility may succeed—or vice versa—despite your best efforts. This can help to greatly simplify the decision-making process for you.

Compatibility, Optimization, and Suitability Considerations

The directive to use only "compatible/optimal/suitable values where appropriate" in step 8 of the "Setting Up a Wireless Profile" section is a classic example of a decision that a wireless network administrator might have to make while configuring a wireless client. Notice, for example, that our WAP supports a lot more authentication and cipher methods than those supported by the wireless client adapter.

So even though the WAP has built-in support for stronger or superior authentication and encryption methods (such as WPA2-Personal/CCMP and WPA2-Enterprise/CCMP) than those supported on the Windows client, the WAP has to be forced to downgrade its own security when communicating with the client to "WPA-Personal/TKIP" simply because the client cannot support anything beyond "WPA-Personal/TKIP." (For more on WPA, WPA2, and TKIP, see Chapter 11.)

The same logic applies to IEEE 802.11 standards used on the wireless client. If the client adapter is not capable of supporting much higher data rates like those offered and supported by the WAP (such as IEEE 802.11n), the AP will be forced to operate at a suboptimal data rate.

The solution to this common mismatch between the client-side devices and the infrastructure-side devices can vary from simple and inexpensive, to complicated and disruptive.

The solution may be as simple as upgrading or replacing the client-side devices with newer hardware that can support the newer standards and functionality. This solution, however, has the disadvantage of breaking support for existing legacy client-side devices (such as devices running older, unsupported OSs), because it may sometimes be impossible to upgrade the legacy client devices. It is possible that the legacy devices or applications may have been originally designed deliberately to function at low data rates and using weak encryption!

Another slightly more complicated and expensive solution may be for the wireless network administrator to try to maintain a wide assortment of infrastructure-side devices to keep everybody happy—the slow clients, the fast clients, the secure clients, and the insecure clients.

Manually Connecting to a Wireless Network

Now let's connect to the wireless network associated with the profile that we created previously.

1. Log into the system.
2. Open the Start menu.
3. In the Search Programs And Files text box, type **connect to a network**. Windows will return a list of items that are related to your search term.

4. From the returned list, click Connect To A Network. A window will appear on the lower-right side of the screen similar to this:

5. Among the list of wireless networks detected, find and select the area-3 network.
6. Click the Connect button. After a moment, Windows will connect to the specified network—if everything goes well.

netsh Utility

netsh is a network configuration utility that ships with newer generations of Windows OSs. And, of course, it comes bundled with Windows 7. It is a command line utility that can be used in performing virtually all the network configuration tasks that can be performed from the Windows graphical user interface (GUI). It can also be used for many advanced tasks, including performing network configuration tasks on remote systems. In addition to allowing you to view and manipulate network configuration, netsh can be used for scripting purposes to help automate common tasks.

Network administrators who are aficionados of the command line interface (CLI) may find that netsh provides a quick way to view and manipulate network settings of a Windows-based computer. This is certainly true if you know your way around the tool and are familiar with its numerous switches and options.

netsh controls certain aspects of the Windows networking stack, which are referred to as *contexts*; these contexts are actually specific network components. Some contexts that are available in Windows 7 are shown in Table 13-3.

We are, of course, mostly interested in the wlan context in this book.

Context	Description
interface	Manages the TCP/IP settings of network interfaces
bridge	Manages network bridge adapters
advfirewall	Manages IPsec, firewall, and other advanced security network components
lan	Manages the wired local area network interfaces
dhcpclient	Manages the DHCP client
wlan	Manages WLAN interfaces
ras	Manages remote access servers

Table 13-3.   netsh Contexts

Wireless Network Configuration Cloning with netsh

Windows makes it easy to clone the wireless configuration settings from one computer to another. This can help speed up provisioning new wireless clients on a network. The automated nature of the cloning process also reduces the chances of human errors when entering network names, secret keys or passphrases, authentication, and cipher values.

Next we will perform a simple configuration clone from one system to another using the powerful netsh command. Start by assuming we have two systems: The primary machine (the source) is the same system that we configured earlier. This machine is named Windows7-A, and it has all the proper wireless settings to connect to our test wireless router, router-W.

The secondary machine (the target) is named Windows7-B. We are going to apply the settings that were cloned from Windows7-A onto this machine.

Perform the following steps on Windows7-A, the source:

1. Log into Windows7-A with a user account with administrative privileges.
2. Choose Start | All Programs | Accessories.
3. Right-click the Command Prompt program and choose Run as Administrator.
4. Select Yes in the User Account Control (UAC) dialog box that opens.
5. A window similar to the one shown next will appear:

6. Type the following command to view the existing wireless profiles stored on the local computer:

```
> netsh wlan show profile
```

7. Then press ENTER.

8. The output should include the wireless profile named area-3 that we created earlier in the chapter.

9. Next, dump (or export) the working profile into a format that can be easily imported into another Windows system to configure it. Use the following command to do this:

```
> netsh wlan export profile name="area-3" folder="C:\Users"
interface="Wireless Network Connection"
```

This command will create a file named Wireless Network Connection-area-3. xml, which will be stored under the C:\Users\ directory. Press ENTER when you're done typing the command.

10. We need to find a way to transport or copy the dumped profile file to the Windows7-B target system—for example, by using a USB flash drive, a CD, a DVD, a network share, or e-mail.

Now perform the following steps on Windows7-B:

1. Log into Windows7-B with a user account with administrative privileges.

2. Choose Start | All Programs | Accessories.

3. Right-click the Command Prompt program and select Run as Administrator.

4. Select Yes in the User Account Control (UAC) dialog box.

5. A command prompt window will appear.

6. Assuming that the wireless configuration profile that was created on the source system (Windows7-A) was copied into a removable media or drive that shows up as drive F: on our target system, we will load (import) the profile from that location. This sample command shows how to do this:

```
> netsh wlan add profile filename="F:\Wireless Network
Connection-area-3.xml"
```

Press ENTER when you're done typing the command.

7. In the window that appears, type the following command to view the newly created wireless profile stored on our target computer:

```
> netsh wlan show profile
```

Press ENTER when you're done typing the command.

8. You should now be able to join and use the same wireless network that the source system (Windows7-A) uses.

TIP   The netsh utility can be used to display useful information about the IEEE 802.11 wireless LAN interface and the device driver information. Among other things, it can also show the 802.11 PHYs supported, and the authentication and ciphers supported. The full netsh command to do this is

```
netsh wlan show drivers
```

Wireless Odds and Ends in Windows 7

Every OS vendor tries to be the first kid on the block to implement support for the latest wireless standards and technologies. Some vendors do a better job than others in this area. Some vendors adopt a wait-and-see approach and wait for the standards or technologies to become mature before incorporating support into their base OSs. In this section, we will take a look at some of the new wireless features that ship with Windows 7.

Virtual WiFi and Software Access Point

Microsoft brings virtualization to the desktop user with the Virtual WiFi (VWiFi) feature in Windows 7. VWiFi uses hardware virtualization techniques to create a virtual instance of the wireless adapter installed in the system. For VWiFi to work, the underlying driver for the adapter has to support this feature; this, therefore, puts some of the onus on the hardware manufacturer to create this support in its drivers. Microsoft has created the framework and necessary hooks into its OS to enable this.

The basic idea behind VWiFi has been around for several years, so most of the major wireless adapter vendors have some support for it, so it should easily work in most newer systems that ship with Windows 7.

VWiFi essentially takes any supported and physically installed wireless adapter and creates a virtual or software version of the adapter. The virtual adapter is directly tied to the main wireless adapter, so if the main wireless adapter is disabled, the virtual adapter is automatically disabled, too.

This virtual adapter shows up as *Microsoft Virtual WiFi Miniport adapter* in the list of network devices in Windows 7 and later OSs.

Due to the natural laws of the universe and physics, *virtualization* is *virtualization* is *virtualization*. This means the original hardware being virtualized will always be the bottleneck in any system; and as such, things cannot run faster than the physical hardware will permit. This is true regardless of what any of the vendor marketing materials may try to tell you. Windows uses some smart internal routines to multiplex the wireless network traffic between the physical and virtual adapters so that associated delays and lags are not so apparent to the end user.

The primary or physical wireless adapter functions in regular wireless STA mode. This means that the adapter can act as a client or member of an infrastructure network and connect to a WAP and perform its normal routines.

The other virtual adapter can, however, function in AP mode, and this is where the name *software access point (SoftAP)* comes from. With the virtual adapter functioning as an AP, other wireless STAs can connect to the SoftAP and form a separate or independent wireless network. This network is the master of its own domain and can have separate security requirements and policies.

The combination of all the new features that VWiFi and SoftAP provide is referred to as a "wireless hosted network."

You can do many cool things with the VWiFi adapter. We look at some of these next.

Internet Connection Sharing

Internet Connect Sharing (ICS) is a feature in modern Windows OSs that facilitates network sharing through a computer running a Windows OS. It essentially turns a computer running Windows into a sort of router or gateway that is capable of performing network address translation (NAT) functions.

Here's how ICS would function in a wireless hosted network scenario:

1. The main node running Windows 7 or later is connected to the Internet via the physical wireless adapter.

2. The main node's access to the Internet is through the physical infrastructure network.

3. The physical wireless adapter can act as a gateway to the VWiFi adapter to provide a way (a route) to reach the Internet.

4. The VWiFi turns into a SoftAP.

5. Because the VWiFi adapter can reach the Internet, the other STAs connected to the SoftAP can also access the Internet, as long as the OS policies allow this.

Configuring SoftAP and VWiFi Using netsh

Windows does not ship with an easy-to-configure point-and-click interface for configuring the wireless hosted network functionality. Setting up SoftAP using VWiFi requires the use of the netsh utility or some other third-party software. We will walk through the steps involved in enabling a wireless hosted network in Windows.

Our sample wireless hosted network will be configured with the parameters shown in Table 13-4.

Perform the following steps on the system running Windows 7 that will host the wireless hosted network:

1. Log into Windows7-A with a user account with administrative privileges.

2. Choose Start | All Programs | Accessories.

Wireless Hosted Network Parameter	Value
SSID	windows-ap
Supported ciphers and authentication	WPA2-Personal CCMP
Key or Passphrase	vwifi-secret
Support IEEE 802.11 standards	Same as those supported by the physical wireless adapter/interface

Table 13-4.   Wireless Client Specifications

3. Right-click the Command Prompt program and select Run as Administrator.

4. Select Yes in the User Account Control (UAC) dialog box.

5. In the window that appears, type the following command to set an SSID of windows-ap for the virtual WiFi interface, and then press ENTER:

```
> netsh wlan set hostednetwork ssid=windows-ap
```

6. Create a passphrase for the wireless hosted network by issuing the following command at command prompt; press ENTER when you're done.

```
> netsh wlan set hostednetwork key=vwifi-secret
```

7. Enable the wireless hosted network by typing this command in the Command Prompt window. Press ENTER when you're done.

```
> netsh wlan set hostednetwork mode=allow
```

8. Start the wireless hosted network by typing this command; press ENTER when you're done.

```
> netsh wlan start hostednetwork
```

9. In the command prompt window, type the following command to view the settings of the wireless hosted network that was just configured:

```
> netsh wlan show hostednetwork
```

10. To view the values of the security parameters used for the wireless hosted network, type this command, and then press ENTER:

```
> netsh wlan show hostednetwork setting=security
```

That's it. The SoftAP should now be set up and broadcasting its SSID.

NOTE The wireless infrastructure network configured here facilitates only local communication; it does not facilitate communication with any external networks such as the Internet. This mode of operation is also known as "standalone mode." Full-mode ICS is needed to enable interconnectivity with external networks. The sidebar titled "Extended SoftAP, VWiFi, and ICS" provides a synopsis of how to configure the ICS component.

You'll perform the following steps on the client system that will connect to the SoftAP. These steps are completely generic and OS-independent, so they can be carried out on almost any wireless client that supports the authentication, cipher, and 802.11 PHYs that SoftAP supports.

1. Use the wireless interface configuration utility for the wireless STA to search for and select the network name (SSID) advertised by the SoftAP. The network should show up as *windows-ap* in the list of available networks.

2. Connect to the windows-ap network.

3. When prompted for the key/passphrase to connect to the SoftAP, enter the key that was created on the SoftAP: **vwifi-secret**.

4. The client STA will become a part of the infrastructure network formed by the Windows-based SoftAP, as long as all the physical (802.11 PHY) and logical (such as Encryption) attributes correlate between the two.

5. The preceding steps can be repeated for other client STAs that want to participate in the wireless hosted network. All the client STAs will be able to communicate with one another as well as with the SoftAP as long as the SoftAP system is up and running.

Note that the steps here will not enable the client STAs to communicate with any external networks, such as the Internet.

Extended SoftAP, VWiFi, and ICS

Here's a quick run-through of the steps involved in configuring full-mode ICS, which will allow the client STAs connected to the SoftAP to access external networks such as the Internet.

The process is two part: The first part involves working from the command prompt, and the second part involves using the GUI tools.

1. Complete the steps in the "Configuring SoftAP and VWiFi Using netsh" section earlier in the chapter.

2. Make sure that system hosting the SoftAP has connectivity to the external Internet via its physical wireless network interface.

3. ICS must be enabled on the physical wireless network interface that has connectivity to the Internet.

4. Choose Start | Search, and type **ncpa.cpl** in the Search Programs And Files text box. Windows will return a list of items that are related to your search term.

5. From the returned list, click the ncpa.cpl program icon to launch the Network Connections Control Panel applet.

6. Locate the network connection that is associated with the physical wireless adapter installed on the system. It might have a name similar to Wireless Network Connection.

7. Right-click the Wireless Network Connection and select Properties. A Wireless Network Connection Properties dialog will appear.

8. Click the Sharing tab in the Wireless Network Connection Properties dialog.

9. Click the check box for the Allow Other Network Users To Connect Through This Computer's Internet Connection option.

10. Under the Home Networking Connection section, click the arrow in the Select A Private Network Connection drop-down list.

11. Select the network connection associated with the virtual WiFi Miniport adapter (VWiFi). It might have a name similar to Wireless Network Connection 2. The final window with all the proper options selected will look similar to the one in this illustration:

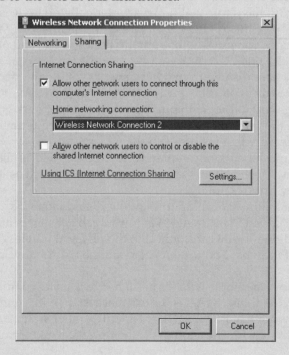

(Continued)

12. Click OK to close the Wireless Network Connection Properties dialog.

13. For good measure, you may want to stop and restart the wireless hosted network service. This can be done by issuing the following commands at the command prompt Press ENTER after each line:

```
> netsh wlan stop hostednetwork
> netsh wlan start hostednetwork
```

14. Issue this command to view the status of the wireless hosted network:

```
> netsh wlan show  hostednetwork
```

15. In case you haven't already done so, perform the four steps on the client system that will connect to the SoftAP, as discussed earlier in the chapter.

From this point, other wireless client STA systems connected to the SoftAP should be able to access external networks such as the Internet.

Mesh Networks

A *mesh network* is a type of network topology in which the nodes work together in harmony to provide a communications link among all nodes. Each node or participant in the mesh network has a role that it plays, but ideally no node is indispensable. The individual nodes have micro-knowledge of their immediate neighbors. This micro-knowledge facilitates communications among nodes at one end of the mesh to other nodes at the extreme end.

Mesh networks are distinct from other network topologies because of their lack of a central or all-knowing controller such as the APs or WAPs found in infrastructure-based wireless networks. The main purpose of a mesh network is not necessarily to provide access to the Internet for the nodes, even though Internet connectivity may be a side benefit or feature that results from the mesh.

The nodes in mesh network can play various roles, such as client roles, router roles, and gateway roles. Mesh networks offer several benefits such as redundancy, cost savings, reliability, and scalability.

The SoftAP functionality built into Windows 7 and later OSs, can be used to build mesh networks. These STAs running Windows 7 can act as repeaters of the wireless radio signals, and one VWiFi can connect to the physical wireless adapter of another STA, whose VWiFi can be used to connect to another STA's physical VWiFi, and so on and so forth.

Figure 13-1 shows a sample wireless mesh network using computers running Windows 7 OS. In the figure, STA-1 is the original system with connectivity to the primary infrastructure network that is managed by AP-1. The other systems, STA-2 and STA-3, have interconnectivity with one another via their physical and virtual wireless interfaces.

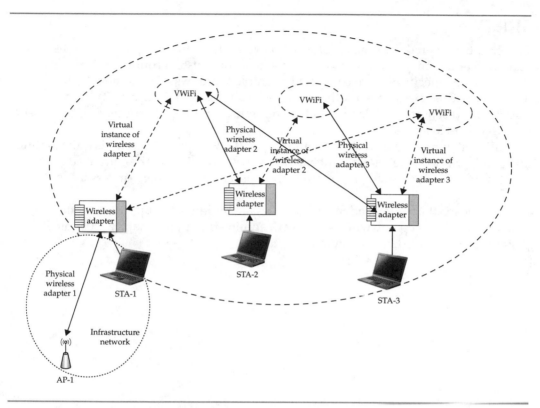

Figure 13-1.   A wireless mesh network

Personal Area Networks (PAN)

Virtual WiFi can be used for linking or connecting miscellaneous types of devices in addition to traditional STAs. This type of connection forms a type of peer-to-peer or ad hoc network. When I say *traditional* STAs here, I am referring to regular mobile or portable computing systems. The other classes of devices (nontraditional) that can take advantage of the VWiFi functionality are consumer electronics devices that are used for personal communications, such as wireless speakers, headsets, phones, printers, PDAs, MP3 players, and so on.

This particular use will become more common place as more wireless device manufacturers implement the new Wi-Fi Direct technology. Wi-Fi Direct helps to reduce the need for traditional wireless infrastructure devices when creating wireless networks to link different types of devices. Wi-Fi Direct can be used to connect various consumer electronics to create a PAN, which can exist independently from the conventional wireless network.

Summary

This chapter covered some basic information you'll need to configure the wireless hardware and management software components of a system running Microsoft's Windows OS. Specifically, we connected a sample Windows client system to the infrastructure wireless network that we judiciously set up in Chapter 12.

We also discussed installation and configuration options and procedures for several wireless client systems.

All in all, wireless network configuration on the Windows platform is quite mature and solid. Windows keeps a decent balance between staying on the cutting edge of wireless technology while being reasonably conservative in other wireless technology areas.

This means that several advanced features are built-in and supported by the OS that should help keep any wireless network administrator busy, happy, frustrated, delighted, entertained, and flustered.

CHAPTER 14 | Apple OS X Clients

Key Skills and Concepts

- Understand the makeup of Macintosh systems.

- Review common Apple wireless networking hardware.

- Configure Apple OS X clients for wireless networking.

- Learn about Internet sharing in the Mac world.

- Review configuration of ad-hoc wireless networks.

- Learn how to configure Bluetooth Dial-up Networking in OS X.

Apple Computers' market share is far from that of the dominant Microsoft Windows, but the number and variety of network-capable devices made by Apple and the fast rate of adoption of these devices make its products an important part of a wireless network administrator's job.

Apple devices come in all shapes and sizes, from servers, to workstations, to all-in-one desktops, laptops, tablets, mobile phones, and portable entertainment units. One thing all these devices have in common is built-in wireless functionality.

Macintosh System Design

Apple manufactures (or brands) the hardware as well as the underlying software that drives their hardware. This has several implications, which we'll examine in the following sections.

Macintosh Software

Configuring and connecting a Machintosh (Mac) system to a wireless network can be a simple process—a testament to why you don't see job advertisements for Macintosh wireless network administrator specialists.

The simplicity offered in Macintosh system has both good and bad sides. The good side is the obvious simplicity and ease of use for the end user.

The disadvantage is a result of the technical sacrifices that had to be made to achieve this simplicity.

Simple systems, by design, generally have less parts and features. It stands to reason that fewer parts and features mean fewer things can go wrong or break. The network or system administrator therefore does not have too much flexibility and configuration options beyond what the Mac designers expose.

Macintosh Hardware

The other facet of Apple's business model that directly affects the administrator's work has to do with the hardware. Mac hardware (and software that drives it) can be very "Apple-centric." Hardware parts and components that are compatible with Apple devices are not as easy to find as commodity PC hardware.

This fact probably has a lot to do with device driver issues, too. If all hardware manufacturers also provided drivers for Macintosh systems, more hardware variety for the consumer would appear on the shelves. But since Apple also manufactures its own hardware, the company probably sees few good business reasons to encourage this.

> **NOTE** A similar issue used to plague the open source software community. Software developers were willing and ready to create drivers for numerous hardware platforms, but hardware manufacturers were resistant to providing details of the inner workings of their hardware to aid the developers. From the hardware manufacturers' perspective, the specific details of the inner workings of their hardware gave the vendors their competitive advantage. This perceived advantage also serves as the market differentiator between similar hardware makers.
>
> This issue hasn't been completely eliminated in the open source software community, but it is vastly improved now.

The Mac system wireless administrator is restricted to a limited range of compatible hardware parts and accessories that Apple makes available, or to those sold by after-market hardware vendors. So, for example, if a wireless administrator needs wireless hardware that implements the latest IEEE 802.11.*xxxxxxx* standard, he or she might be out of luck until Apple releases such a hardware or a firmware update that implements the new functionality, or until some of the few after-market vendors offer the hardware.

Macintosh Wireless Hardware

Almost everything wireless produced by Apple likely includes the word "AirPort" in its name. The current generation of wireless adapters for Macintosh systems has names like these:

- **AirPort** This was used to identify Apple's first-generation wireless hardware that implemented the original IEEE 802.11b standard. This AirPort hardware operates in the 2.4 GHz wireless band.

  Even though it may be technically inaccurate, this is still the generic name of most modern Macintosh wireless hardware.

- **AirPort Extreme** This generation of hardware is the successor of the original AirPort cards. The AirPort Extreme hardware implements the next-generation wireless standards—the IEEE 802.11a/g/n. Among other improvements, this generation offers better data rates and security. Within the AirPort Extreme family is the AirPort Extreme 802.11n hardware, which specifically implements support for the IEEE 802.11n standard. The most current AirPort Extreme hardware can operate in the dual 2.4 and 5 GHz wireless frequency bands; as a result, AirPort Extreme is backward-compatible with the first-generation AirPort wireless networks.

- **AirPort Extreme Base Station** This is Apple's implementation of a wireless access point (WAP). Underneath the hood of the Extreme base station are the typical electronics and radios that are found in other vendors' residential access points (APs). But as with most things Apple, the base station has been wrapped up in a pretty package.

Current versions of the base station can operate in dual-band mode, which means that they support both frequencies in the 2.4 and 5 GHz range. It supports the IEEE 802.11a/b/g/n wireless standards.

■ **Time Capsule** This multipurpose wireless device is primarily a wireless network storage device that can also function as WAP. In other words, it's an AirPort Extreme Base Station with a big hard drive inside it.

The network storage component of time capsule works in conjunction with software installed on a compatible Mac computer to perform automatic data backups from the client system to the time capsule.

The current generations of the time capsule hardware are multi-band devices that can operate in the 2.4 and 5 GHz frequency bands.

Mac OS X Wireless Client Configuration

In this section, we will connect a Mac OS X computer to a wireless network. Our sample system will be configured to connect to the generic WAP that we set up in Chapter 12.

Table 14-1 shows the properties of our WAP, and Table 14-2 shows the properties of our Macintosh wireless client STA.

Setting Up the Hardware

Out of the box, Macintosh systems already have the appropriate drivers installed for the hardware, so there's really nothing for you to do. The rest of this chapter assumes

Specification	Value
AP host name	wap-2
SSID	area-2
Supported ciphers and authentication	WEP 64 / 128 bits WPA-Personal / TKIP WPA-Personal / CCMP WPA-Enterprise / TKIP WPA2-Personal / TKIP WPA2-Personal / CCMP WPA2-Enterprise / CCMP
Encryption key or passphrase	!never-never-land2!
Wireless PHYs supported	802.11b, 802.11g, 802.11a, 802.11n

Table 14-1. WAP Specifications

Specification	Value
Host name	book01
Operating system	Mac OS X - Snow Leopard
Wireless card vendor and type	Apple / AirPort Extreme
Wireless card chipset/driver	Atheros
Supported ciphers and authentication	WEP WPA-Personal / TKIP WPA-Personal / CCMP WPA-Enterprise / TKIP WPA2-Personal / TKIP WPA2-Personal / CCMP WPA2-Enterprise / CCMP
Compatible/optimal/suitable cipher and authentication	WPA2-Personal
Supported IEEE 802.11 standards	802.11b, 802.11g
Compatible/optimal/suitable IEEE 802.11 standard	802.11g

Table 14-2. Wireless Mac Client Specifications

that the wireless adapter on the Macintosh client system is the one that shipped with the system.

Setting Up a Wireless Location

Other mainstream operating systems (OSs) associate different network configurations with *profiles*. The equivalent of profiles in the Mac world is a "location." On a Mac, a set of network settings and different devices are associated with a location. A defined location allows the ordering of the various connectivity devices and interfaces within it. For example, at a location named *home*, wireless connectivity would be preferred over wired connectivity; and at a separate location called *office*, Ethernet connectivity is preferred over wireless. A default location called "Automatic" exists on Macintosh systems, and we'll use this default location in our sample system.

Mac OS X offers several ways of connecting to a wireless network: it can be done automatically or you can do it manually. Allowing OS X to configure a system to connect to a wireless network can be easier for you, but in certain scenarios it can also lead to unintended consequences, such as connecting to an unknown or hostile wireless network. We will opt for the manual approach here, because it allows us better control of the process and offers some insight into what OS X is doing in the background.

AirPort Utility

Macs ship with a built-in AirPort utility that can be accessed by choosing Applications | Utilities.

It is easy to assume that the utility is used for managing the AirPort wireless adapter built into the computer. However, the AirPort utility is used to manage a separate proprietary hardware sold and marketed by Apple called the AirPort.

This AirPort device is nothing more than a standalone WAP or residential gateway device that can be used to create simple wireless networks for sharing Internet access and other resources. Wireless networks created and controlled by AirPort base stations are, however, not proprietary. The networks are accessible by regular Microsoft Windows clients, Linux/UNIX/BSD clients, and other IEEE 802.11-compliant devices.

Like other management interfaces for configuring and managing wireless infrastructure devices, the AirPort utility can be used for configuring Dynamic Host Configuration Protocol (DHCP) options, radio settings, access controls, authentication, and encryption settings for the AP.

Manually Connecting to a Wireless Network

Let's connect to the wireless network associated with the AP that was configured in Chapter 12. The SSID of that network is area-2.

1. Log into the system.
2. Click the Apple menu at the upper-left corner of the screen and choose System Preferences. A System Preferences window similar to the following appears:

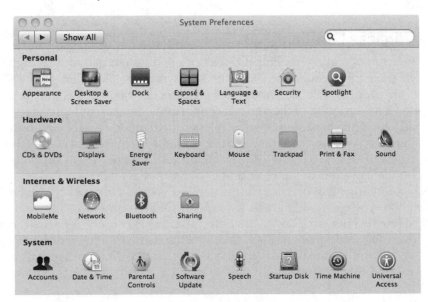

3. Click the Network icon under the Internet & Wireless category. The Network preferences window appears, showing the available network interfaces/ devices on the system in the left pane.

4. Click the AirPort network device in the left pane. The network window will display some of the configuration options available for the AirPort adapter.

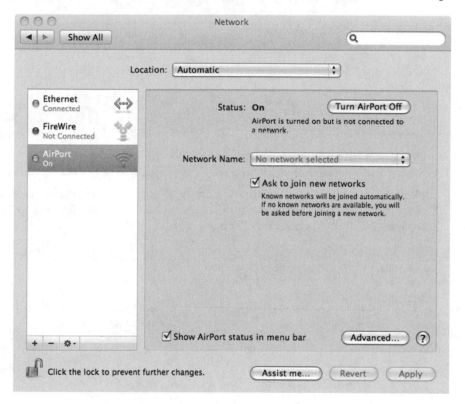

5. In the AirPort network screen, make sure that the AirPort device status is set to On. If the status is set to Off, click the Turn AirPort On button.

6. Click the Advanced button at the bottom of the window. An AirPort configuration screen will appear with a list of preferred networks (if any).

7. Click the plus (+) sign under the Preferred Networks section. The following dialog box will appear, where you can input the settings for the wireless network:

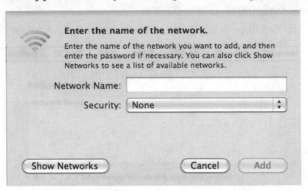

8. Enter the SSID in the Network Name field. Click the arrows in the Security dropdown list and select the security type used on our sample network (see Table 14-1).

9. Click the Add button after completing all the fields. You will be returned to the AirPort network configuration window. Clear the Remember Networks This Computer Has Joined check box. The window should look like this:

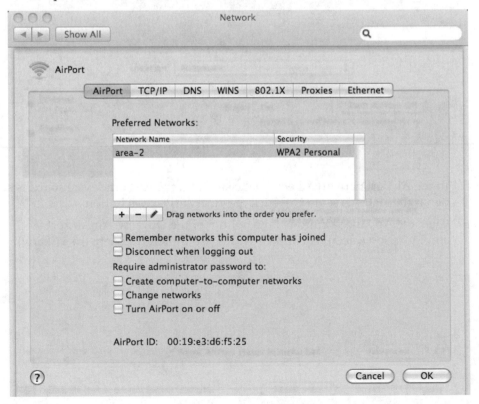

10. Click OK to close the AirPort network window.

11. Click Apply in the main Network window to initiate and complete the client's wireless connection with the infrastructure wireless network (area-2).

12. Close the system preferences window.

Changing Connection Preferences

In Mac OS X systems, the default ordering for network interfaces in a given location is Ethernet, Firewire, and AirPort. You can change the default order by following these steps:

1. While logged into the system, click the Apple menu at the upper-left corner of the screen and select System Preferences. The System Preferences window will appear.

2. Click the Network icon under the Internet & Wireless category. The Network preferences window will appear, showing a list of the available network interfaces in the left pane.

3. In the bottom-left corner of the window, click the arrow next to the little flower or sun-shaped icon and select Set Service Order from the drop-down list. A Service Order dialog will appear:

Service Order:

```
Ethernet
FireWire
AirPort
```

Drag services to change order.

Cancel   OK

4. Click any of the available interfaces and drag it to the desired position.

5. Click the OK button when you are done.

6. Click the Apply button in the Network preferences window and then close the window.

Wireless Odds and Ends in OS X

Most OS vendors try to be the first to implement support for the latest wireless standards and technologies. Other vendors wait for the standards or technologies to become mature before incorporating support into their base OSs.

Mac OS X falls into the conservative category. Apart from making it easy to configure network settings, the OS does not offer many extra bells and whistles in the way of advanced features for the network administrator. But it does offer a few extra wireless networking functionalities. We'll take a brief look at a few of these.

Internet Sharing

Macs have their own implementation of Internet connection sharing that allows an OS X system to become an Internet gateway of sorts to other systems on a network. This functionality, however, is not quite as polished or feature-rich as that of other popular mainstream OSs, such as FOSS/Linux-based systems or Windows systems.

Internet sharing on Macintosh systems works like this:

1. The Mac connects to the Internet via any of the interfaces available on the system—AirPort card, Ethernet, Bluetooth, dial-up modem, and so on.

2. Computers that need to access the Internet connect via the Mac host using any of the *other* ports that are *not* being used by the Mac to connect to the Internet. The key word here is *other*.

3. So, for example, if the Mac host is connected to the Internet via an Ethernet connection, other systems can access the Internet via the Mac's wireless AirPort interface.

The Macintosh's AirPort connection is shared by other connected computers via an Ethernet Local Area Network (ELAN). To share an Internet connection over Ethernet, the Mac needs to be connected to the Internet by some other means, such as AirPort, Bluetooth, cellular network, and so on. The converse is also true: To share an Internet

It's All About the Sharing

Internet connection sharing is called *Internet sharing* in the Macintosh world. In this world, we don't share the connection, instead we share the Internet. *Sharing* refers to accessing and using network and other system resources provided by other Mac systems. So, for example, if you want to turn a Mac system into a web server, you'd turn on the Web Sharing service.

So instead of saying we want to "enable the complicated routing, Network Address Translation/Port Address Translation (NAT/PAT) firewall gateway functions" in a Macintosh system, we can simply say we want to "enable Internet sharing."

connection over the AirPort interface, the Mac needs to be connected to the Internet by some other means, such as wired Ethernet, Bluetooth, cellular network, and so on.

Let's walk through the steps for enabling Internet sharing on our sample Macintosh system. The network we will set up will resemble the one shown in Figure 14-1.

1. Click the Apple menu at the upper-left corner of the screen and select System Preferences.

2. In the System Preferences window, click the Network icon under the Internet & Wireless category. The Network preferences window will appear.

3. Click the AirPort network device. You'll see some of the configuration options available for the AirPort adapter.

4. Make sure that the AirPort device is turned on. The Status should read "Connected" and should also show that AirPort is connected to an infrastructure network (for example, area-2). If the status is set to Off, click the Turn AirPort On button.

5. After making sure that the AirPort is properly connected to a network, click the Show All button at the top of the window to return to the main System Preferences window.

6. Click the Sharing icon under the Internet & Wireless category. The Sharing window will appear.

7. In the left pane, look through the list of available services and click the Internet Sharing service. The Internet Sharing configuration options will appear on the right side of the window.

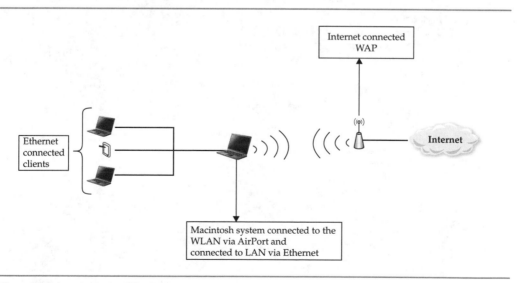

Figure 14-1. Internet Sharing

8. Click the drop-down arrow in the Share Your Connection From field and select the AirPort option.

9. In the To Computers Using field, click the Ethernet checkbox to enable it.

10. Make sure that the Internet Sharing service is still highlighted in the left pane of the window. Click the checkbox next to Internet Sharing to start the service. A caution screen will pop up, asking if you are sure about turning on Internet Sharing. Click the Start button.

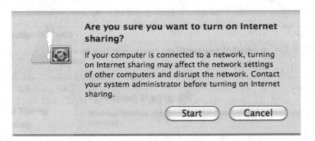

11. The final Sharing preferences window will resemble the one shown here. Close this window when you've finished making changes.

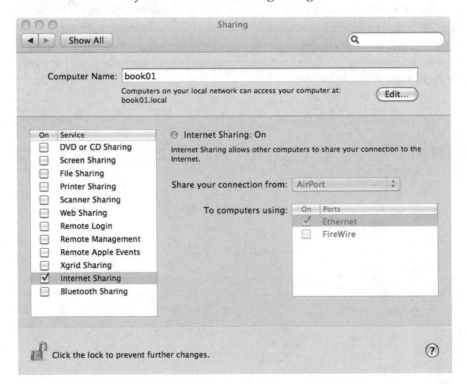

All systems that are connected to the same Ethernet segment (via a switch or directly) to the Mac will automatically get IP configuration information from the Mac system. All systems connected via the Mac system should now be able to access any networks (such as the Internet) that the Mac system can access.

Ad-Hoc Networks

Apple's OS makes it easy to set up ad-hoc networks (also known as computer-to-computer) networks. Ad-hoc networks make creating a wireless network of systems quick work, without the need for a traditional infrastructure device such as an AP.

Unfortunately, the ad-hoc networks that can be created by Mac systems cannot be protected with any of the newer high-grade encryption technologies. At the time of this writing, only 40-bit and 128-bit WEP can be used in securing ad-hoc networks.

Ad-hoc networking and traditional infrastructure-based WLANs are mutually exclusive in Mac computers. Enabling and hosting an ad-hoc wireless network on a Mac disconnects the system from any other infrastructure networks.

We'll create our ad-hoc network with the configuration settings shown in Table 14-3. Use the following steps to set up an ad-hoc network on a Mac system:

1. Click the Apple menu at the upper-left corner of the screen and choose System Preferences. Click on the Network icon under the Internet & Wireless category.

2. In the Network preferences window, click the AirPort network device to see some of the configuration options available for the AirPort adapter.

3. Make sure that the AirPort device Status is set to On. Click the Turn AirPort On button if the status is set to Off.

4. In the Network Name field, click the drop-down arrow and select Create Network at the bottom of the list. A Create a Computer-to-Computer Network window will appear.

5. By default, the Network Name (SSID) field will show the host name of the computer. Change the name (SSID) to **ad-hoc-mac**.

Specification	Value
SSID/network name	ad-hoc-mac
Channel	Accept the default value
Password	128-mac-adhoc
Security	128-bit WEP

Table 14-3. Ad-Hoc Network Settings

6. Click the Require Password checkbox to enable that option. Complete the rest of the fields with the values from Table 14-3 and shown here:

7. Click the OK button to close the window. Then click Apple in the Network preferences window.

8. Close the Network preferences window.

The ad-hoc network will now be active and ready for other wireless clients to join it.

Mac Ad-Hoc Network Considerations

The process for connecting to ad-hoc networks is not much different from connecting to regular infrastructure wireless networks. The same tools used for connecting to infrastructure networks are used. Theoretically, you should just locate the network name (SSID) in the list of detected networks and input any security parameters, and the WLAN client should be good to go.

However, because ad-hoc networking is not covered by any IEEE 802.11 standard, your mileage may vary when connecting different clients from other non-Apple platforms to a Macintosh computer–hosted ad-hoc network.

The wireless clients participating in a Mac-hosted ad-hoc network use dynamic-link-local addresses (aka self-assigned IP address). These link-local addresses are in the 169.254.0.0/16 range and are covered in the IETF document RFC 3927.

Bluetooth Dial-up Network

The Dial-Up Network (DUN) Bluetooth profile allows a Bluetooth device to be used as a data modem with other compatible Bluetooth devices such as laptops and personal digital assistants (PDAs). Connecting (or *teetering*) to Bluetooth mobile devices uses the cellular data network and as such can incur additional charges beyond the regular voice plans offered by cellular network providers.

 NOTE   Don't forget to disconnect the connection properly when you're done with this process to avoid unnecessary charges from the mobile network operator.

Here's how to set up a Bluetooth DUN:

1. Click the Apple menu and select System Preferences.
2. Click the Network icon under the Internet & Wireless category. The Network preferences window will appear.
3. In the left pane, click the plus (+) sign at the bottom left to add a new network interface type.
4. In the dialog box that appears, click the Interface drop-down list and choose the Bluetooth DUN interface type. Accept the default Service name of Bluetooth DUN in the Service Name field. The completed dialog box will look like this:

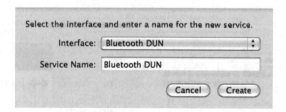

5. Click the Create button. Back in the main Network preferences window, click the Apply button.
6. Click the newly created Bluetooth DUN interface in the left pane.
7. Click the Set Up Bluetooth Device button.
8. Make sure that the Bluetooth-enabled mobile device is powered on and set in a discoverable mode.

9. In the Bluetooth Setup Assistant window, the Mac system will search for and display a list of Bluetooth-enabled devices in the vicinity. When the desired device has been discovered, highlight the device in the list of devices (see the illustration) and then click Continue.

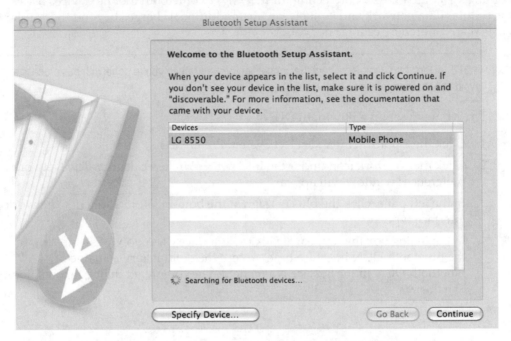

10. A passkey will be automatically generated by the Macintosh system to complete the pairing process with the target Bluetooth device (such as a mobile phone). The generated passkey needs to be entered on the target Bluetooth device exactly as displayed in the Macintosh Bluetooth Setup Assistant window (see the illustration). Return to the Macintosh after entering the passkey for the target device and click Continue.

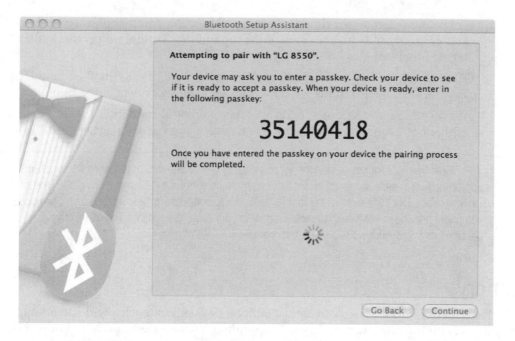

11. The Bluetooth Setup Assistant process will continue to the Bluetooth Mobile Phone Setup stage. This is where things get personal and you will supply the values that apply to your particular mobile or cellular provider. Click Continue when you are done. On our sample system, we use the following values:

12. A final Conclusion window will appear, informing you that pairing was successful. You'll also see some of the capabilities (supported profiles) of the target Bluetooth device. Click Quit.

13. Back at the main Network preferences window, make sure the Bluetooth DUN interface is selected in the left pane, and then click Connect in the right pane.

14. If everything checks out, the Macintosh system will connect to the Internet via a Bluetooth connection to the mobile device using the cellular network.

TIP   Feel free to experiment with the list of available phone vendors or mobile/cellular network operators, just in case the exact provider for your phone is not listed (see step 12 of the Bluetooth DUN section). Because most of the mobile network providers use similar underlying technologies, you might find that if you select a phone vendor in your area, the settings and connection parameters might work with your specific provider.

Summary

It takes a lot of effort to design an almost foolproof system for such a wide demographic of people, and the engineers at Apple Corporation have created a system that works right out of the box—as long as you don't plan on doing too much beyond what the engineers intended the system to do. A Mac network doesn't leave much work for a wireless network administrator—no room, for example, to show off or demonstrate any advanced wireless networking expertise.

CHAPTER 15 | Linux Clients

Key Skills and Concepts

- Learn about wireless hardware on Linux platforms.
- Learn about the Linux Wireless Extensions application programming interface (API).
- Learn about some popular Linux wireless drivers.
- Review proprietary wireless firmware in Linux-based distributions.
- Review the tools used for managing wireless settings in Linux clients.
- Walk through configuring a wireless Linux client using GUI tools.
- Walk through configuring a wireless Linux client using command line tools.
- Learn how to set up an ad-hoc wireless network using Linux clients.

Gone are the days when installing and configuring new hardware in Linux-based systems required excessive preparation, endless cups of coffee, studying README files and other documentation, compiling and cross-compiling, reverse-engineering, researching help forums, and muttering unprintable words. These days, it's pretty much plug and play.

Linux is at the heart of most Free and Open Source Software (FOSS) distributions, and as such, Linux is responsible for controlling the interactions between hardware components and the other parts of the operating system (OS).

Thousands of Linux-based distributions are available today. The distributions are backed by different commercial and noncommercial entities and hobbyists. They pull together different FOSS projects and applications and package them into a distribution format. The objective of most distributions is to produce an easy to install, easy to use, easy to distribute, multipurpose or single-purpose set of software.

One of the great advantages of FOSS is that it gives you the freedom to be as hands-on or as hands-off as you like. You can get down and dirty with the working details, or you can choose to point-and-click from predefined menu options to get the job done. We'll use a hands-on approach in this chapter and provide more information than is necessary to configure a Linux-based client for wireless networking. Hopefully, you'll gain some insight into how things work internally, to help you glue the theoretical components of wireless networking with the practical aspects so that you can be adept at troubleshooting and marketable as a wireless network administrator.

The concepts and fundamentals discussed in this chapter can also be applied to the inner workings of other OSs; the difference is in how other closed systems might force you to take a hands-off approach to learning and implementing.

Wireless Hardware on Linux Platforms

Wireless hardware used to be especially tricky to get working on Linux-based systems for several reasons. Open source hardware developers were willing and able to write drivers for the wireless hardware, but hardware vendors were not forthcoming with the details of the inner workings of their hardware to support this. Some device manufacturers thought that revealing the inner workings of their hardware meant giving up their market advantage to the competition.

Fear, uncertainty, and doubt (FUD) have also played a part. Various countries have designated local legislative bodies that allocate and manage the use of the radio spectrum within the country. This is important for organizational, safety, and practical purposes, for example, so that amateur radio enthusiasts don't transmit their messages on the same frequency channels used by local emergency services.

Some wireless hardware vendors exhibit FUD characteristics because they believe that Linux drivers cannot adhere to the requirements for radio spectrum use. The FUD factor comes into play because some hardware vendors cannot see any way for the open source developer community to enforce the rules because of the very nature of FOSS. Such vendors believed it would be easy for end users to alter underlying code that enforced the rules that force compliance with local radio spectrum use.

All FUD aside, the FOSS community has gone to great lengths in recent times to reach out and assuage the concerns of wireless hardware vendors. Practical safety measures now included in the Linux kernel code ensure that the code can't be changed in unacceptable ways.

Linux Wireless Drivers

A common joke in the FOSS community is that any single problem can be solved in 300 different ways. This joke belies the good and bad aspects of FOSS.

The obvious good part is the choice and power that FOSS gives the users. The bad aspect is the confusion, disjointedness, and seeming incoherence that can result from having so many approaches and solutions. But die-hard FOSS purists point out that the competition that this breeds is healthy and that eventually only the very best of the solutions (software projects) will survive and last.

FOSS drivers for wireless hardware are affected by issues similar to those that affect regular software applications. This means that, in some instances, multiple device drivers are available for the same wireless hardware family. Sometimes, multiple application programming interfaces (APIs) exist in the Linux kernel space for drivers to use.

Mainstream FOSS wireless driver families and APIs will be discussed, beginning with the existing API families.

Linux Wireless API

A wireless API in the Linux world provides a standard way to manipulate wireless networking hardware. The API presents a device driver–independent method of implementing the 802.11 standard in the Linux kernel. APIs provide device driver developers a standard set of tools to use for communicating with the Linux kernel so that they don't have to start from scratch when adding standard functions. Some of the popular API families are discussed next.

Wireless Extensions API

The Wireless Extensions API represents one of the earliest attempts at creating a wireless API for Linux and is possibly the most popular and widely deployed API. Wireless Extensions has been ingrained in the Linux kernel since version 2.0.03.

Wireless Extensions APIs are no longer under active development. Even though new features are no longer being added, newly discovered bugs are still being fixed because of the large user base that depends on the API.

The Wireless Extensions API comprises three parts:

- ■ **User interface**   Userland tools for manipulating certain wireless hardware settings (see Table 15-1)
- ■ **Kernel interface**   The Linux kernel resident parts of the Wireless Extensions API
- ■ **Driver interface**   Specific hooks in the hardware device drivers that developers must include in the driver code to enable them to take advantage of the benefits of the extensions

Mac80211

Mac80211 is the Linux wireless API du jour. It is the wireless subsystem in the Linux kernel. The most current work in the Linux wireless API space is in developing the mac80211 software stack, which is being groomed as the successor to the Wireless Extensions API.

Wireless Extensions Utility	Description
iwconfig	Used primarily for configuring certain wireless characteristics of wireless network interfaces, such as the frequency or channel, operating mode, transmit power, sensitivity, modulation, extended service set ID, to name a few. Also used for displaying basic information about the wireless interface.
iwlist	Used for querying and displaying extended information about wireless network interfaces.
iwevent	Used for debugging. It displays wireless events generated by drivers and settings changes.

Table 15-1.   Wireless Extensions API Userland Tools

In its current state, mac80211 is geared toward *SoftMAC* devices, which are network devices that depend on software or firmware to implement a part of their functionality. Mac80211 therefore provides the framework with which developers can write drivers for SoftMAC wireless devices.

Like its predecessor (Wireless Extensions), mac80211 comprises several components:

- **cfg80211**   This is the new Linux wireless configuration API component of the mac80211 framework. In addition to configuration, cfg80211 also helps with registering the wireless subsystem (mac80211) with the networking subsystem of the Linux kernel.

  An important function of cfg80211 is in addressing the regulatory compliance concerns of wireless hardware vendors (discussed in the section "Wireless Hardware on Linux Platforms"). Out of the box, wireless hardware can support different radio frequency channels, but cfg80211 handles compliance issues by ensuring that the wireless hardware operates only in the allowed radio channels for the currently set regulatory area.

- **nl80211**   The user-space tools interact with cfg80211 via the nl80211 interface in the new mac80211 framework.

- **iw**   A set of command line interface (CLI) userland tools for querying and configuring the interface and radio properties of a wireless network device. iw offers similar functionality to that offered by the iwconfig utility under the Wireless Extensions API.

Common Linux Wireless Drivers

There are countless wonderful open source wireless drivers available today, but we'll only discuss the popular driver families and their associated chipsets. Remember that only a handful of vendors manufacture wireless chipsets, and everybody else integrates those chipsets into their products. As a result, you will find that one driver family is able to support multiple wireless hardware from seemingly different vendors.

One of the unique attributes of the FOSS world is that, instead of the actual device manufacturers spearheading the process of shipping and maintaining the drivers for their products, the FOSS community often takes the initiative to do this. You will therefore find some wireless hardware drivers that are part of open source projects.

MadWifi Driver Project

The MadWifi—Multiband Atheros driver for WiFi—family supports wireless devices with chipsets manufactured by Atheros Corporation. The Atheros chipsets are found in numerous wireless devices all over the world, including products manufactured by Apple, Netgear, Toshiba, HP, Linksys, and D-Link, to mention a few.

Some of the drivers to emerge from the MadWifi project, such as the following, are not considered fully open source, because a part of their functionality is implemented in a closed source binary format. The closed source part is implemented as a hardware abstraction layer (HAL). Despite this fact, Wi-Fi hardware sporting the Atheros chipsets are popular in the FOSS community, because they offer a wide range of advanced features and are easy to extend. Atheros is also regarded as a FOSS-friendly company.

- **madwifi**   The original set of drivers from the MadWifi project. These drivers rely on the HAL component to function, and although they are no longer under active development, they still have a large install and user base.

- **ath5k**   This driver is tagged as the replacement for the original madwifi driver. Unlike madwifi, ath5k does not depend on a closed source HAL component to function. This therefore makes it a completely open source driver. The driver supports Atheros wireless chipset types with model numbers of AR5*xxxx*.

- **ath9k**   This is the Atheros driver for the next-generation wireless chipsets that supports the IEEE 802.11n standard. It is also a completely FOSS driver since it does not require a closed source, binary-only HAL to function. The driver supports Atheros wireless chipset types with model numbers of AR9*xxxx* and some AR5*xxxx* models.

Broadcom Drivers

Broadcom is a popular manufacturer of the chipsets used by various OEMs in their finished products—such as Acer, Apple, Asus, Belkin, Buffalo, Dell, HP, Microsoft, Linksys, and USRobotics, to mention a few.

The Broadcom chipsets make use of firmware. (See "Firmware and HAL" for more on firmware.)

Broadcom wireless chipset–based wireless devices have a high success rate for working in FOSS platforms, even though Broadcom as a company has not been helpful in driver development. As a matter of fact, most of the Broadcom drivers were developed by reverse-engineering efforts by members of the FOSS community. In fact, wireless devices with USB interfaces that sport Broadcom chipsets are notoriously known to be unusable with FOSS.

Following are two of the drivers for wireless chipsets made by Broadcom:

- **b43legacy and bcm43xx**   The b43legacy driver is considered as a legacy driver for older hardware and older kernels. Specifically, b43legacy supports wireless hardware that implements only the IEEE 802.11b standard and other so-called revision 2 chipsets. bcm43xx is a much older driver that is no longer used in newer Linux kernels.

- **b43**   This is the current driver that is used for driving newer Broadcom wireless chipsets and hardware.

Intel Wireless Drivers (iwlwifi Project)

Intel wireless drivers are mostly released and managed as open source software projects, supported by Intel. However, the chipsets themselves often require a closed source firmware component. The most recent wireless drivers for Intel hardware comes packaged with recent Linux kernels and FOSS distributions. The official project name under which all recent Intel wireless driver development takes place is called *iwlwifi*, and the newer iwlwifi drivers make use of the new mac80211 wireless API subsystem.

Following are some of the drivers for wireless chipsets made by Intel:

- **ipwXXXX** This driver family is used by early generation wireless networking chipsets made by Intel, such as those in Intel PRO/Wireless 2100, 2200, and 2915 hardware series. This driver family does not make use of the new mac80211 API.

- **iwl3945** This driver is for the Intel PRO/Wireless 3945ABG/BG family of wireless hardware chipsets and replaces the older ipw3945 driver for the same hardware family.

- **iwlagn** This driver is for the Intel Wireless WiFi Link AGN hardware family, such as 5100BG, 5100ABG, 5100AGN, 6000AGN, 4965AGN, and so on. It supports the newer generation wireless chipsets that implement the IEEE 802.11n standard.

Firmware and HAL

Firmware is a type of software that runs directly on the hardware (microcontroller, or the digital signal processor) for which it is written. Any hardware designed to use firmware needs the firmware for its basic functionality. The firmware's role is not usually that of enforcing policy issues or protecting the host system in which it is being used.

The HAL is also a type of software, but it is intended to be executed on the central processing unit (CPU) of the host like other regular system software. As implied in its name, the HAL provides a layer of abstraction between the hardware and other system processes. This abstraction or protective layer can be used, for example, to make sure that a wireless radio does not operate in radio frequencies that are noncompliant with requirements or legislation in a certain area.

Open source OS distributions cannot legally package proprietary firmware for chipsets along with other distribution software, because the firmware is often copyrighted by the device manufacturer. Closed source OSs such as Microsoft Windows and MAC OS X do not have these issues, because the hardware manufactures often package the device firmware with the device drivers that are sold or shipped with the finished products.

Some of the mainstream distributions work around this proprietary firmware limitation by creating simple methods for users to legally "obtain" the firmware from legal sources (the hardware vendor). One such popular mechanism for obtaining the firmware is by using the b43-fwcutter utility.

Because Windows drivers for wireless hardware can be easily downloaded from the vendor's web site, b43-fwcutter can be used to extract the firmware packaged with the Windows driver.

(Continued)

The extracted firmware is then placed in a specific location on the file system, so that it can be loaded or executed by the chipset hardware.

This is the current workaround as of this writing. It is possible that things can change in the future for better or for worse, such as the following:

- Wireless hardware manufacturers may find ways (legal or technical) to prevent tools such as b43-fwcutter from functioning.

- Wireless hardware manufactures may decide to start directly supporting FOSS users better and create drivers with or without firmware for FOSS-based distributions.

Wireless Network Management Tools

Every OS has its own set of tools or utilities to aid the user in configuring the wireless network devices on the host. Linux-based distributions are no exception. And, as you may have come to expect of FOSS platforms, not only do we have one tool, but we have several tools from which to choose.

The FOSS community started with a lot of different wireless configuration tools, and as the laws of natural selection would have it, only a handful have survived and gained popular acceptance. The GUI tools used for configuring wireless devices in Linux are referred to collectively as Wireless Network Managers. And the command line tools are referred to as, you guessed it, CLI utilities. Incidentally, most of the nice GUI network managers still call and depend on the CLI utilities in their back-ends!

We'll briefly look at some of the wireless configuration tools and managers next.

NetworkManager

This GUI-based network management application is possibly the most popular and widely deployed of all the management tools.

Over the years, it has become so neatly integrated with the rest of the graphical desktop experience in FOSS distributions that it is easy to forget it is even there! The design goals of nm include ease of use, visual clarity, and a motto that things should "just work" with as little interaction as possible.

NetworkManager, or nm, is used to manage various types of network communication interfaces, such as mobile broadband, Wi-Fi, modems, virtual private networks (VPNs), and so on. It does this via a plug-in architecture that makes it future-proof, because as new networking protocols and interface types emerge, plug-ins can be developed to integrate with nm.

nm expects that the relevant drivers for the underlying hardware are already functional. It does not concern itself with such things and leaves these low-level details to the OS. So buyer-beware, because if you run out and purchase the latest and greatest wireless network adapter that implements the nonexistent IEEE 802.11h.u.g,e standard, nm will not necessarily help you. However, it will help as long as the drivers exist for the card and are usable by the kernel.

Wireless Interface Connection Daemon

Wireless Interface Connection Daemon (wicd, pronounced "wicked") is an alternative open source wired and wireless network management application for FOSS-based distributions. It also has its own large user following, but its install base is not as large as that of nm. It has a few outstanding features:

- Offers seamless integration with all the different desktop environments
- Has no dependencies on the popular GNOME desktop environment
- Offers a powerful command line console interface

More information about wicd can be found at the project web site at http://wicd .sourceforge.net.

ConnMan

This is the newest kid on the block in the FOSS network management tools world. ConnMan's focus is a little different from that of the other major players in network management space. The project's focus is on managing Internet connection settings in embedded devices running Linux.

ConnMan is especially suited for embedded and low-powered devices because of its purported small footprint, which means that it tries to use as few system resources as possible.

Like the other network management tools, ConnMan uses a plug-in architecture so that it can be easily extended to support new and emerging networking technologies and protocols.

NOTE   Keep in mind that ConnMan's stated focus is the same type of modest focus (or ambitions) that everything starts with. There are too many real-life examples of companies or projects that start off as small with a niche focus and end up everywhere! So don't be surprised if you start seeing ConnMan replacing established applications such as nm on the desktop platforms in the coming years.

Good Ol' CLI

When all the nice GUI network management tools fail, there will always be the CLI. It was hinted earlier that the GUI network management tools rely on the command line tools in one form or another in their back-ends, so having a good understanding of how to use the CLI tools is never a waste and can come in handy in many situations—such as when you're troubleshooting or debugging wireless connectivity issues, for testing purposes, or even when the GUI tools are simply not available.

Some of the CLI tools that can be used for configuring wireless network interface in FOSS systems are described in the following sections.

iwconfig

iwconfig is used for configuring the wireless characteristics of wireless network interfaces. It can be used for querying and manipulating Open System Interconnection (OSI) layer 1 and layer 2 properties of network devices.

NOTE   See the "Wireless Extensions API" section earlier in the chapter for more information.

The syntax for iwconfig is shown in here:

```
Usage: iwconfig [interface]
               interface essid {NNN|any|on|off}
               interface mode {managed|ad-hoc|master|...}
               interface freq N.NNN[k|M|G]
               interface channel N
               interface bit {N[k|M|G]|auto|fixed}
               interface rate {N[k|M|G]|auto|fixed}
               interface enc {NNNN-NNNN|off}
               interface key {NNNN-NNNN|off}
               interface power {period N|timeout N|saving N|off}
               interface nickname NNN
               interface nwid {NN|on|off}
               interface ap {N|off|auto}
               interface txpower {NmW|NdBm|off|auto}
               interface sens N
               interface retry {limit N|lifetime N}
               interface rts {N|auto|fixed|off}
               interface frag {N|auto|fixed|off}
               interface modulation {11g|11a|CCK|OFDMg|...}
```

To learn more about the syntax and various options that can be used with the `iwconfig` command, consult built-in help system (often referred to as the "man page"). To see more documentation about `iwconfig`, type the following command at the shell prompt:

```
man iwconfig
```

The functionality provided by the `iwconfig` command is being replaced by the `iw` command, discussed next.

iw

This is the new command line wireless configuration utility for wireless devices in FOSS systems. It makes use of the nl80211 interface of the new mac80211 wireless API in the Linux kernel.

Like iwconfig, iw can also be used for querying and manipulating OSI Physical layer (layer 1) and the Data Link layer (layer 2) properties of network devices.

The syntax for using iw and some of the common options are shown here:

```
Usage: iw [COMMAND]
Where COMMAND can be any combination of the following:
Commands:
      event [-t] [-f]
      phy
      list
      dev <devname> set channel <channel> [HT20|HT40+|HT40-]
      dev <devname> set freq <freq> [HT20|HT40+|HT40-]
      dev <devname> set type <type>
      dev <devname> set meshid <meshid>
      dev <devname> info
      dev <devname> del
      dev <devname> interface add <name> type <type>
      dev <devname> ibss join <SSID> <freq in MHz> [fixed-freq] [<fixed bssid>]
      dev <devname> ibss leave
      dev <devname> station dump
      dev <devname> station set <MAC address> plink_action <open|block>
      dev <devname> station del <MAC address>
      dev <devname> station get <MAC address>
      dev <devname> get mesh_param <param>
      dev <devname> set mesh_param <param> <value>
      dev <devname> scan [-u] [freq <freq>*] [ssid <ssid>*|passive]
      dev <devname> scan dump [-u]
      dev <devname> scan trigger [freq <freq>*] [ssid <ssid>*|passive]
      reg get
      reg set <ISO/IEC 3166-1 alpha2>
```

Table 15-2 shows some sample iw command line options and usage.

To learn more about the syntax and various options that can be used with the iw command, consult the man page. To see more documentation about the iw command, type the following at the shell prompt:

```
man iw
```

wpa_supplicant

This is the Swiss army knife equivalent of a wireless security configuration tool in the FOSS world. It is also available on other OS platforms. Among other things, wpa_supplicant can be used for configuring the authentication, authorization, association, and encryption parameters of a wireless STA.

Specifically, it is a Wi-Fi Protected Access (WPA) client and an IEEE 802.1X supplicant. It is responsible for negotiating encryption keys and parameters with an authenticator

iw Command and Options	Description
iw list	List all wireless devices and their capabilities.
iw event	Monitor wireless events from the kernel. Useful when debugging authentication, deauthentication, association, and disassociation issues.
iw dev wlan0 station dump	Show wireless station statistic information for the wireless interface wlan0.
iw reg get	Display the kernel's current regulatory domain information.
iw reg set <ISO/IEC 3166-1 alpha2>	Change the current regulatory domain.

Table 15-2.   iw Toolset

(such as a wireless access point) and is also responsible for negotiating Extensible Authentication Protocol (EAP) authentication parameters with the authentication server (such as a RADIUS server). To be pedantic, wpa_supplicant implements the WPA Supplicant component of the IEEE 802.11i standard.

The syntax and options that can be used with the wpa_supplicant program are shown here:

```
Usage: wpa_supplicant [OPTIONS]
Where some of the common OPTIONS can be any combination of the following:
-B = run daemon in the background
-c = Configuration file
-i = wireless interface name
-d = increase debugging verbosity (-dd even more)
-D = driver name (can be multiple drivers: nl80211,wext)
-f = log output to debug file instead of stdout
-g = global ctrl_interface
-K = include keys (passwords, etc.) in debug output
-t = include timestamp in debug messages
-p = driver parameters
-P = PID file
-q = decrease debugging verbosity (-qq even less)
-u = enable DBus control interface
-N = start describing new interface
```

To learn more about the syntax and various options that can be used with the wpa_
supplicant command, consult the man page. To see more documentation about the
wpa_supplicant command, type the following command at the shell prompt:

```
man wpa_supplicant
```

wpa_passphrase

This utility is an accoutrement to the wpa_supplicant toolkit. wpa_passphrase takes an
ASCII input and generates WPA preshared key, using the supplied service set identifier
(SSID) as the *salt* (random information added to a password to make extracting the
password more difficult).

Wireless clients use the resulting key to encrypt the network traffic.

The output generated by wpa_passphrase can be used to create an important stanza
in wpa_supplicant's main configuration file (wpa_supplicant.conf). This stanza is used
to begin new wireless network configurations.

The syntax and options for wpa_passphrase are straightforward:

```
wpa_passphrase usage:
wpa_passphrase <ssid> [passphrase]
```

For example, given a passphrase of "sample-passphrase" and an SSID of "sample-
ssid," we can use wpa_passphrase to generate a preshared key and a wireless network
configuration stanza like the one here:

```
$ wpa_passphrase  sample-ssid   sample-passphrase
network={
     ssid="sample-ssid"
     #psk="sample-passphrase"
     psk=531c39e23b7791bc6fd02eb1376f34d77de09b827a167c676f47f6a0b5a4d957
}
```

To learn more about the syntax and various options that can be used with the wpa_
passphrase command, consult the man page. To see more documentation about the
wpa_passphrase command, type the following at the shell prompt:

```
man wpa_passphrase
```

nm-tool

This is a nifty little tool. From its name, you can tell that it is related to the GUI
NetworkManager software.

nm-tool provides information and other wireless statistics about NetworkManager,
wireless devices, and wireless networks.

It does not accept any command line options and its usage is dead simple. At the
command line, just type this:

```
nm-tool
```

ifconfig

ifconfig is used for configuring the network interfaces on Linux-based systems. It is used for querying and manipulating OSI Network layer (layer 3) and the Transport layer (layer 4) properties of network devices. ifconfig is therefore not used for configuring the radio or IEEE 802.11 characteristics of wireless interfaces. That function is left for iwconfig and iw to do.

Once the radio properties for a wireless adapter have been set up, ifconfig can perform the relevant network protocol–level configurations. It can be used for setting up protocol-level parameters such as the TCP/IP configuration for the wireless (or wired) interface.

The following syntax and options can be used with ifconfig:

```
Usage: ifconfig device address [options]
```

Where *device* is the name of the Ethernet device (for instance, eth0), *address* is the IP address you want to apply to the device, and *options* are one of the following:

Option	Description
Up	Enables the device. This option is implicit.
Down	Disables the device.
Arp	Enables this device to answer arp requests (default).
-arp	Disables this device from answering arp requests.
mtu *value*	Sets the maximum transmission unit (MTU) of the device to *value*. Under Ethernet, this defaults to 1500.
netmask *address*	Sets the netmask to this interface to *address*. If a value is not supplied, ifconfig calculates the netmask from the class of the IP address. A class A address gets a netmask of 255.0.0.0, class B gets 255.255.0.0, and class C gets 255.255.255.0.
broadcast *address*	Sets the broadcast address to this interface to *address*. If a value is not supplied, ifconfig calculates the broadcast address from the class of the IP address in a similar manner to netmask.
pointtopoint *address*	Sets up a point-to-point connection (PPP) where the remote address is *address*.

To learn more about the syntax and various options that can be used with the ifconfig command, consult built-in man page. To see more documentation about the ifconfig command, type the following at the shell prompt:

```
man ifconfig
```

ip

Another powerful program that can be used to manage network devices in Linux is the ip program. This utility comes with the iproute software package, which contains networking utilities (such as ip) that are designed to use the advanced networking capabilities of the Linux kernel. The syntax for the ip utility is a little more terse and less forgiving than that of the ifconfig utility.

ip is the preferred toolkit for configuring the network interfaces on newer Linux-based systems. It provides similar functionality to that provided by ifconfig but is much more powerful than ifconfig.

It is used for querying and manipulating OSI Network layer (layer 3) and Transport layer (layer 4) properties of network devices.

ip does not concern itself with the radio or IEEE 802.11 characteristics of an interface. That function is left for iwconfig and iw to do. Once the radio properties for a wireless adapter have been set up, ip can perform the relevant network or transport level configurations.

To learn more about the syntax and various options that can be used with the ip command, consult built-in man page. To see more documentation about the ip command, type the following at the shell prompt:

```
man ip
```

Ubuntu Wireless Client Configuration: nm

Ubuntu is one the most popular and highly regarded Linux-based distributions. It is backed and sponsored by Canonical Ltd.

Ubuntu is a derivative of another popular Linux distribution called Debian. As in other popular Linux-based distributions, Ubuntu is an amalgamation of numerous open source projects and the Linux kernel.

We'll configure a sample system running the Ubuntu OS to connect to one of the WAPs that we configured in Chapter 12. Some of the relevant properties of the WAP are listed in Tables 15-3, and Table 15-4 shows the properties of our Ubuntu client acting as our wireless client STA as well as some of our configuration objectives.

1. Log into the Ubuntu system as a regular user.

2. Launch the NetworkManager GUI tool to set up the wireless interface using any of these methods to open the Network Connections window shown in the following illustration:

   - Simultaneously press ALT-F2 to launch the Run Application dialog box. Type **nm-connection-editor** in the text box and click the Run button.

   - Locate the NetworkManager system tray applet in the top-right corner of the desktop. Right-click the icon and select the Edit Connections option.

Specification	Value
Host name	wap-1
SSID	area-1
Supported ciphers and authentication	WEP 64 / 128 bits WPA-Personal / TKIP WPA-Personal / CCMP WPA-Enterprise / TKIP WPA2-Personal / TKIP WPA2-Personal / CCMP WPA2-Enterprise / CCM
Encryption key or passphrase	$never-never-land1$
Wireless PHYs supported	802.11b, 802.11g, 802.11a, 802.11n

Table 15-3. WAP Specifications

Specification	Value
Host name	foss01
Operating system	Ubuntu Linux
Wireless card vendor and type	Atheros
Wireless card driver	ath5k
Supported ciphers and authentication	WEP WPA-Personal / TKIP WPA-Personal / CCMP WPA-Enterprise / TKIP WPA2-Personal / TKIP WPA2-Personal / CCMP WPA2-Enterprise / CCMP
Compatible/optimal/suitable cipher and authentication	WPA2-Personal / CCMP
Supported IEEE 802.11 standards	802.11b, 802.11g, 802.11a, 802.11n
Compatible/optimal/suitable IEEE 802.11 standard	802.11n
Network name or SSID	area-1

Table 15-4. Ubuntu Client Wireless Specifications

■ Click System in the Main Menu area located in the top-left corner of the desktop. Select Preferences and click Network Connections in the program group.

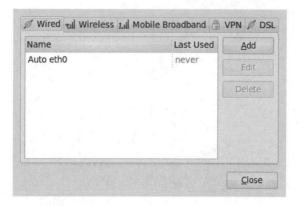

3. Click the Wireless tab in the Network Connections window.

4. Click the Add button. A new dialog box will appear, where you can set up the parameters for connecting to the wireless network.

5. Complete the fields with the information from Tables 15-3 and 15-4 so that the completed screen resembles this:

6. Click the Wireless Security tab. Complete the fields with the information in Tables 15-3 and 15-4 so that the completed screen resembles this:

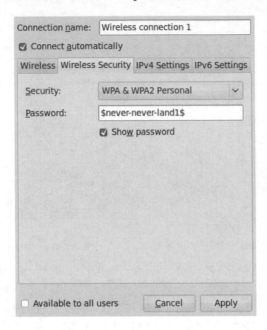

7. Accept the default settings in the IPv4 Settings and IPv6 Settings tabs.

8. Click the Apply button.

9. The passphrase (secret key) that was entered in step 6 is stored in a *keyring* so that the system remembers it for you in the future. A keyring dialog window might open, prompting you about allowing the NetworkManager applet access to the newly created Wireless network. Click Always Allow.

10. The NetworkManager applet will immediately attempt to connect to the WAP using the parameters just supplied. And if all goes well, you should see a brief notification message (shown next) in the upper-right area of the desktop, informing you that the connection to area-1 has been established.

11. The ensuring Network Connections configuration window will resemble the one shown next. Click the Close button to exit the NetworkManager program.

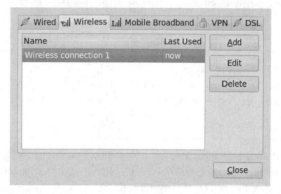

 TIP The NetworkManager has a nifty little applet in the upper-right area of the system tray (see step 2). Clicking this applet can provide useful information, such as a quick visual indication of radio signal strength, the wireless network name to which the system is currently connected, and other wireless networks that the Wi-Fi adapter detects in the vicinity.

Figure 15-1 shows an example of some of the information that the system tray applet displays on our sample system.

Figure 15-1. Information displayed by the system tray applet

Generic Linux Wireless Client Configuration: CLI

Here we'll configure a sample system running the Ubuntu OS to connect to one of the WAPs we configured in Chapter 12.

Some of the relevant properties of the WAP are shown in Table 15-3. Table 15-4 shows the properties of our Linux client acting as our wireless client STA, as well as some of our configuration objectives.

The configuration will be using the CLI of our sample system. We will use some of the utilities and programs discussed earlier—iw, wpa_supplicant, and wpa_passphrase.

NOTE   The commands and utilities we'll use for the command line configurations will need to be executed from a shell prompt or a terminal. We therefore need a terminal program or a terminal emulator program of some sort. As with everything FOSS, you have several options, some of which might already be installed by default. Some of the popular terminal program choices are xterm, gnome-terminal, and konsole. We will use xterm on our sample system.

In addition, the following procedures are lengthy, because extra information has been interspersed with the steps in the process. This extra information might prove useful to extra-curious readers. The little extras are completely optional and should not affect our final objective. The information is marked with an [OPTIONAL] tag. Feel free to skip steps that begin with [OPTIONAL] if you are so inclined.

PHY and MAC Layer Configuration: CLI

Let's start by configuring the PHY and MAC layers using the CLI:

1. Log into the Linux system as a regular user.

2. Press ALT-F2 keys to launch the Run Application dialog box.

3. Type **xterm** in the text box and click the Run button. A xterm window similar to the following will appear. This is our first terminal window.

```
username@foss01:~$
```

4. Some of the commands we will need to configure the wireless settings manually need to be executed by a user with administrative privileges on the system—such as the root user. If you are already logged into the system as the root user, you can ignore this step and jump to step 5.

On Debian-based systems such as Ubuntu, you can temporarily become the root user by typing (the command to type begins after the $):

```
username@foss01:~$ sudo su -
```

Then enter the appropriate password when prompted.

On Red Hat–based systems (such as Fedora, Red Hat Enterprise Linux [RHEL], openSUSE, and Centos), you can temporarily become the root user by typing this command (shown in boldface):

```
username@foss01:~$ su -
```

Enter the appropriate password when prompted.

5. The NetworkManager applet is usually running quietly in the background to help manage the network devices in newer Linux distributions. Since we want to do things manually here, we will temporarily stop the nm service so that it does not get in our way. To stop the service, type the following:

```
root@foss01:~$ service network-manager  stop
```

6. [OPTIONAL] We need to create a simple configuration file for the wpa_ supplicant utility, which will contain the name of the wireless network with which we want to associate, as well as the secret key for the WPA2-PSK encryption type we are using. We will use wpa_passphrase to generate a sample output. The output will be printed directly to the terminal. This is just a test run. Type the text shown in boldface:

```
root@foss01:~$ wpa_passphrase  area-1  '$never-never-land1$'
    network={
    ssid="area-1"
    #psk="$never-never-land1$"
    psk=6ababa9040d5f450e0ad7221febead68a3008863fac6e9eee364478a4
615c6ec
    }
```

7. The test output looks fine, so we can go ahead and write the output directly to our sample configuration file. The file will be named area-1-wpa.conf. Type this:

```
root@foss01:~$ wpa_passphrase  area-1  '$never-never-land1$'  >
area-1-wpa.conf
```

We are almost ready to connect to our WAP.

8. [OPTIONAL] Before we connect, let's find a way to monitor some of the stuff going on in the background when a wireless STA connects to an access point (AP). To do this, we launch another terminal and run different commands there to watch what happens.

9. [OPTIONAL] Repeat steps 2 and 3 to launch a new xterm window—our second terminal window. Arrange the first and second xterm windows on your screen so that you can view them side-by-side without one overlapping the other.

10. [OPTIONAL] We will use the `iw` command with its `event` option to display wireless events in *real time* as they are generated by the wireless adapter driver and other changes.

    On Debian/Ubuntu systems, install iw if you don't have it installed already, by typing

    sudo apt-get install iw

    On Red Hat–based systems, install `iw` by typing **yum install iw**.

    This command should be typed in the second xterm window:

    ```
    root@foss01:~$ iw event
    ```

11. Use wpa_supplicant to connect to the WAP over the interface wlan0, using the configuration file we created in step 7. Type this:

    ```
    root@foss01:~$ wpa_supplicant -Dnl80211 -iwlan0 -c ./area-1-wpa
    .conf
    CTRL-EVENT-SCAN-RESULTS
    WPS-AP-AVAILABLE
    Trying to associate with 00:14:6c:dd:4a:1c (SSID='area-1'
    freq=2442 MHz)
    Associated with 00:14:6c:dd:4a:1c
    WPA: Key negotiation completed with 00:14:6c:dd:4a:1c [PTK=TKIP
    GTK=TKIP]
    CTRL-EVENT-CONNECTED - Connection to 00:14:6c:dd:4a:1c completed
    (auth) [id=0 id_str=]
    ```

12. [OPTIONAL] The console of the second xterm window will be updated as new wireless events are detected. It might have an output similar to this on our sample system:

    ```
    root@foss01:~$ iw event
    wlan0 (phy #0): scan finished
    wlan0 (phy #0): auth 00:14:6c:dd:4a:1c -> 00:16:ce:6a:2e:ba
    status: 0: Successful
    wlan0 (phy #0): assoc 00:14:6c:dd:4a:1c -> 00:16:ce:6a:2e:ba
    status: 0: Successful
    ```

    Our Linux-based wireless STA should now be properly authenticated and associated with the WAP. This means that everything is okay at the PHY and MAC layers between the STA and the AP. This is the same as wireless or radio link.

Network and Transport Configuration: CLI

Now we'll deal with the network and transport layers of the communication link so that we can actually do something useful with it:

1. Repeat steps 2 and 3 (in the "PHY and MAC Layer Configuration: CLI" section) to launch a new xterm window. This will be our *third* xterm window. Arrange the first, second, and third xterm windows on your screen so that you can view them without one overlapping another.

   This part of the configuration process should be easy, because, thankfully, we have a Dynamic Host Configuration Protocol (DHCP) server running on wap-1 (the AP). We configured the DHCP server service earlier on in Chapter 12 when we configured our infrastructure devices.

2. Most FOSS distributions ship with DHCP client functionality. One popular FOSS implementation of DHCP client software is the dhclient, which is used for configuring interfaces using DHCP. Run the dhclient utility with the name of the wireless interface (wlan0, in our case) as the only argument. Type this:

```
root@foss01:~$ dhclient  wlan0
Internet Systems Consortium DHCP Client
For info, please visit http://www.isc.org/sw/dhcp/
Listening on LPF/wlan0/00:16:ce:6a:2e:ba
Sending on   LPF/wlan0/00:16:ce:6a:2e:ba
Sending on   Socket/fallback
DHCPDISCOVER on wlan0 to 255.255.255.255 port 67 interval 4
DHCPOFFER of 172.16.1.3 from 172.16.1.254
DHCPREQUEST of 172.16.1.3 on wlan0 to 255.255.255.255 port 67
DHCPACK of 172.16.1.3 from 172.16.1.254
bound to 172.16.1.3 -- renewal in 39788 seconds.
```

   According to the output, it looks like we are good to go. We were able to get an IP address lease from wap-1. The new IP address of the Linux client STA is 172.16.1.3. And the IP address of the DHCP server (wap-1) is 172.16.1.254.

3. Let's do some simple diagnostic tests and gather some statistics for the wireless connection to make sure that we have IP layer connectivity with wap-1. Type this:

```
root@foss01:~$ ping -c 2 172.16.1.254
PING 172.16.1.254 (172.16.1.254) 56(84) bytes of data.
64 bytes from 172.16.1.254: icmp_seq=1 ttl=64 time=1.18 ms
64 bytes from 172.16.1.254: icmp_seq=2 ttl=64 time=4.72 ms
--- 172.16.1.254 ping statistics ---
2 packets transmitted, 2 received, 0% packet loss, time 1001ms
rtt min/avg/max/mdev = 1.189/2.956/4.724/1.768 ms
```

   Looks good.

4. Now let's view some of the interface statistics. Type this:

```
root@foss01:~$ iw dev wlan0 station dump
Station 00:14:6c:dd:4a:1c (on wlan0)
        inactive time:   3572 ms
        rx bytes:        3534066
        rx packets:      27233
        tx bytes:        163800
        tx packets:      1281
        signal:          -12 dBm
        tx bitrate:      54.0 MBit/s
```

5. Display the PHY capabilities of the wireless devices:

```
root@foss01:~$ iw list
Wiphy phy0
        Band 1:
                Frequencies:
                        * 2412 MHz [1] (20.0 dBm)
                        * 2417 MHz [2] (20.0 dBm)
                    ...<TRUNCATED>....
                        * 2484 MHz [14] (20.0 dBm)
(passive scanning)
                Bitrates:
                        * 1.0 Mbps

                        * 2.0 Mbps (short preamble supported)
                    ...<TRUNCATED>....
                        * 54.0 Mbps
        Band 2:
                Frequencies:
                        * 5180 MHz [36] (30.0 dBm)
(passive scanning, no IBSS)
                        * 5200 MHz [40] (30.0 dBm)
(passive scanning, no IBSS)
                    ...<TRUNCATED>....
                Bitrates:
                        * 6.0 Mbps
                        * 9.0 Mbps
                    ...<TRUNCATED>....
                        * 48.0 Mbps
```

6. Display the PHY layer and MAC layer information about the wireless interface:

```
root@foss01:~$ iw dev wlan0 scan dump
BSS 00:14:6c:dd:4a:1c (on wlan0)
        TSF: 9180742729 usec (0d, 02:33:00)
        freq: 2422
        beacon interval: 100
        capability: ESS Privacy ShortPreamble ShortSlotTime
(0x0431)
        signal: -23.00 dBm
        SSID: area-1
        Supported rates: 1.0* 2.0* 5.5* 11.0* 6.0 9.0 12.0 18.0
        DS Paramater set: channel 3
        ERP: <no flags>
        Extended supported rates: 24.0 36.0 48.0 54.0
        WPA:    * Version: 1
                * Group cipher: TKIP
                * Pairwise ciphers: TKIP
                * Authentication suites: PSK
        WMM: parameter: 01 80 00 03 a4 00 00 27 a4 00 00 42 43 5e
00 62 32 2f 00
```

That's it.

Using Propriety Drivers or Firmware for Wireless Adapters

The following procedures show how a sample Ubuntu system deals with detecting and installing propriety software that cannot be legally packaged with the OS. I assume that the system used here has a connection to the Internet through some interface other than the wireless interface, since the wireless interface is currently disabled due to lack of proper driver or firmware.

1. While logged into the system, make sure that the system has a connection to the Internet via some other means, such as a wired connection (Ethernet).

2. Press ALT-F2 to launch the Run Application dialog box.

3. Type **jockey-gtk** in the text box and click the Run button.

4. The utility will launch and start by searching for any relevant hardware on the system that may need proprietary drivers or firmware to function.

(Continued)

Our sample system uses a wireless network adapter with a Broadcom chipset, which needs some proprietary software to function. A hardware drivers screen appears:

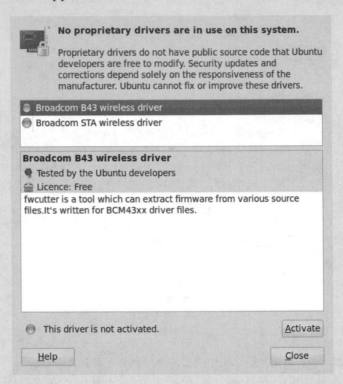

Select the appropriate driver for your hardware from the list, and then click the Activate button. For our sample system, we choose the open source B43 driver combined with the closed source firmware. The b43-fwcutter utility will take care of the proprietary firmware for us.

5. A dialog box might open, prompting you to authenticate to the system before being permitted to perform the privileged operation. Enter the necessary password in the Password text box and click Authenticate.

6. The program will proceed to downloading and installing the necessary software for the hardware. After completion, the hardware drivers screen will change to reflect the fact that proprietary software is in use

on the system. The screen will resemble the one shown next. Click the Close button to exit the application.

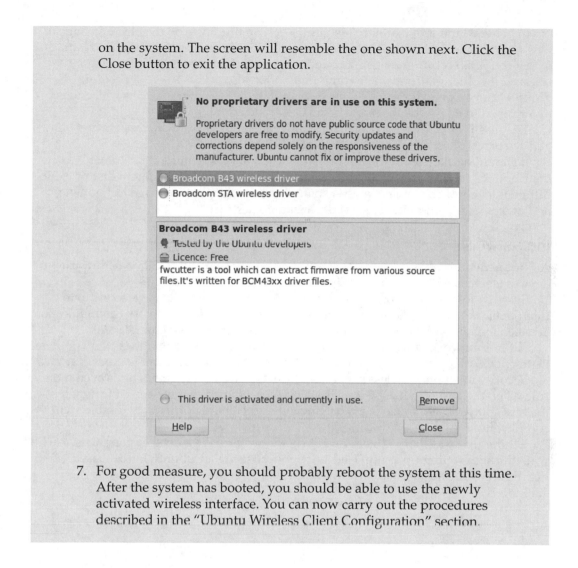

7. For good measure, you should probably reboot the system at this time. After the system has booted, you should be able to use the newly activated wireless interface. You can now carry out the procedures described in the "Ubuntu Wireless Client Configuration" section.

Wireless Odds and Ends in FOSS/Linux

As with the other mainstream OS platforms discussed so far in this book, FOSS-based systems have their own implementations of various new technologies in the wireless networking world.

As mentioned in the preceding chapters, every OS vendor tries to be the first to implement support for the latest wireless standards and technologies. Other vendors wait for the standards or technologies to become mature before incorporating support into their base OSs. Linux and FOSS–based systems are no exception.

In general, and as a matter of fact, the FOSS community is on the bleeding edge of new technologies, which has its advantages and disadvantages. One of the advantages is that the FOSS community gets a chance to play with new toys before they are released on other platforms. The community often also acts as a sort of proofing ground (guinea pigs) for some of this new stuff. If the technology works well and is determined to be useful, the mainstream OSs try to adopt it.

The disadvantage to living on the cutting edge of things is that you get what you get. There are often no guarantees or warranties of any sort.

It should be clarified at this point that within the FOSS community, numerous distributions (distros) target different users. This means that the different distros exhibit different levels of conservativeness—some ship with very cutting edge technologies and others ship with more conservative technologies.

Ad-Hoc Networks

Ad-hoc networks are wireless networks that do not make use of traditional infrastructure devices (such as APs) for connectivity among wireless stations (STAs).

Systems connected to an ad-hoc network do not automatically have access to the Internet via the host. An ad-hoc network is used only to facilitate network communication between members of the network. These systems are on their own little islands.

Ad-hoc networks are not covered by any official standards, but they are very popular. Different OS platforms have their own implementations with different features.

Security in ad-hoc wireless networks is usually a toss of the coin. This is even more true when the ad-hoc networks consists of STAs running different OSs. Unfortunately, the best compatibility mode when dealing with different platforms is to disable security completely.

Setting up an ad-hoc network on FOSS platforms can be a dead easy process. We'll walk through setting up a simple ad-hoc network using an Ubuntu Linux–based distribution. The ad-hoc network will be created with the configuration settings in Table 15-5.

Specification	Value
SSID/ network name	ad-hoc-linux
Channel	Accept the default value
Passphrase	128-linux-adhoc
Wireless security	WEP 128-bits

Table 15-5. Ad-Hoc Network Settings

Perform the following steps on the system that will be initiating and hosting the ad-hoc network—aka, the source:

1. Log into the Ubuntu system as a regular user.

2. Locate the NetworkManager system tray applet in the top-right corner of the desktop. Click the icon and select the Create New Wireless Network... option. A New Wireless Network dialog box will appear, which looks similar to this:

3. Complete the fields of the dialog box with the information from Table 15-4. The completed screen will look like this:

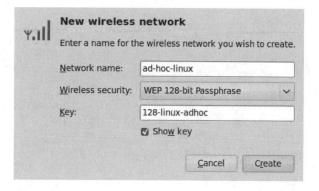

4. Click the Create button.

5. The NetworkManager applet will immediately attempt to enable the new ad-hoc network on the local system. If all goes well, you should see a brief message in the upper-right area of the desktop informing you that the ad-hoc-linux connection has been established.

6. The ad-hoc network with the SSID of ad-hoc-linux will now be active and ready for other wireless clients to join it.

dnsmasq, iptables, dhcp, and dns in Ad-Hoc Networks

Several cool things happen "automagically" in the background whenever a system is configured to host an ad-hoc network in a Linux-based system.

One of these things relies on the dnsmasq package (application), a caching DNS proxy and DHCP server with a very small footprint. dnsmasq is installed by default in recent versions of Ubuntu, but you can quickly install it on older systems if necessary (using `sudo apt-get install dnsmasq-base`).

The program ships with a sensible set of defaults so that it just works without any user input. For example, on an Ubuntu/Debian-based system, the following is the default behavior of dnsmasq:

1. Assign the network interface of the host managing dnsmasq the IP address of 10.42.43.1.

2. Assign DHCP address to clients in the range 10.42.43.10 to 10.42.43.100.

3. Assign the clients with a gateway address of 10.42.43.1.

These defaults mean that all stations connected to the ad-hoc network will get assigned IP addresses in the range of 10.42.43.10 to 10.42.43.100, and the host running the show will be assigned an IP address of 10.42.43.1. This is how the systems on the ad-hoc network get their TCP/IP configuration. And the defaults can be changed by explicitly running dnsmasq with a configuration file that contains the desired parameters.

The second thing that happens is that the packet filter subsystem of the Linux kernel gets activated. That subsystem is known as netfilter, and the utility that manages that subsystem is known as iptables. iptables is used for managing packet forwarding, routing, firewall, and packet logging functions in the Linux kernel.

iptables can also have different front ends (GUI, scripts) to make it easier to use. One such front end is the Uncomplicated Firewall (ufw) in Ubuntu/Debian-based systems.

iptables is used to create simple firewall rules to protect the host managing the ad-hoc network.

Once the ad-hoc network is disabled, the automatic firewall rules and other settings created by dnsmasq are removed, and the dnsmasq daemon goes back to sleep.

Perform the following steps on the other systems that will join the ad-hoc network (the target systems):

1. Use any wireless network management utility available on the system to locate and connect to the ad-hoc network with the SSID of ad-hoc-linux.

   - In other Linux-based systems, you can use nm, the command line, wicd, or ConnMan.

   - In Mac OS X systems, you can use the network preferences application.

   - In Microsoft Windows systems, you can use the Manage Wireless Networks Control Panel applet.

2. Supply the WEP passphrase from Table 15-5 when prompted for a password or key to join the network.

If everything goes well, the system will be connected to the ad-hoc network and will be able to communicate with other systems that are connected to the same ad-hoc network.

Summary

The Linux OS is at the heart of most Free and Open Source Software (FOSS) distributions. In this chapter, you learned about wireless hardware compatibility issues on FOSS platforms, and you learned about Linux wireless drivers and the different Linux wireless APIs. You learned how to use GUI and command line tools to set up and configure hardware and firmware for wireless Linux-based client systems. You also learned how to set up an ad-hoc wireless network for Linux clients.

Relative to the preceding chapters discussing the other mainstream platforms (Windows and Apple OS X), a lot of extra information and detail was provided in this chapter. The very nature of FOSS automatically gave us the choice, the resources, and the tools to communicate this level of detail.

CHAPTER 16 | Plan, Design, Survey, and Deploy

Key Skills and Concepts

- Review and reveal the wireless network administrator's Quad-W rule of thumb.
- Understand the importance of planning when designing wireless networks.
- Learn the art and importance of wireless site surveys.
- Review the special considerations for different wireless network deployment scenarios.

Because some of what you'll be reading about in this chapter will make sense only after you've read the information in earlier chapters, we placed the cart before the horse by setting up our infrastructure devices and connecting our wireless client stations (STAs) to the network. In the real world, we should formally plan and scope out our wireless network before throwing the hardware into it. So we will do what we should have done four chapters ago here.

In the real world, before deploying hardware on a wireless network, a wireless network administrator should follow the *Quad-W (WWWW)* rule of thumb. The Quad-W rule is simple: wireless network planning, wireless site survey, wireless network design, and wireless network deployment. We start by planning the network using information gathered during the site survey. Certain elements and information gathered from the site survey activities show up in the planning and designing stages. As such, the site survey goes hand in hand with the planning and designing activities. Next, the information garnered from the site survey is used to tweak and optimize the network design. Finally, the actual deployment can begin.

The discussions in this chapter take a high-level view and include a few low-level technical details of planning, designing, and deploying a wireless network.

Wireless Network Planning and Designing Considerations

A properly planned wireless network supported by a solid network design and implementation process can help detect and prevent issues that can become a source of headaches in the future. We'll start by taking a look at network hardware.

Past, Present, and Future Hardware

The incredible growth and fast rate of development in today's wireless industry make it easy for wireless network administrators and users to end up with obsolete or unsupported hardware before they know it. In some respects, this particular issue is a little tricky to prevent, because nobody has a crystal ball to predict the future.

But wireless network administrators can do a few things to help ensure that they get the most out of their hardware purchases. Plan your network to accommodate existing legacy hardware and technologies—within reason, of course. For example, it doesn't make any sense to force all the wireless stations on the network to run at IEEE 802.11 speeds just because one station does not support any of the IEEE 802.11 b, g, or n standards. In this case, it would be best to bite the bullet and upgrade the single legacy client to support the newer wireless PHYs so that all the clients can reap the benefits of the newer PHY standards.

You should also plan to include the hardware and technologies available today. This is a no-brainer, since you probably don't have much of a choice in this regard. Only hardware that is readily available can be used in deploying a wireless network.

Finally, you should plan to use the hardware and technologies that are currently under development. These hardware and technologies will be adopted in the near future. Consider, for example, the recently ratified IEEE 802.11n standard. This standard had been in the draft stage for a long time, but hardware vendors were already manufacturing numerous products that supported the draft standard for several years before its release.

In this case, it would have been quite prudent and safe to assume that the IEEE 802.11n would be *the* wireless LAN standard of the future and plan accordingly. Practical and prudent planning in this case would ensure that any new wireless equipment acquisitions had some support for the draft standard, or at least an easy, built-in upgrade path to the standard.

Interoperability and Incompatibility Issues

Before recommending a bunch of expensive infrastructure hardware for purchase, a wireless network administrator must make sure that existing client hardware will work properly with the new hardware to avoid interoperability and incompatibility issues. These issues can occur in hardware products whose manufacturers do not design and build their hardware to conform to universally accepted standards. Original equipment manufacturers (OEMs) sometimes do not fully adhere to standards in their product designs, hoping to gain market advantages over their competition with "innovative" unique designs.

Consider, for example, a wireless OEM that advertises a *turbo-super-duper* operation mode that promises never-before-seen WLAN speeds. The devices are able to achieve these speed improvements through nonstandards-based and proprietary means. The effect of this, however, is that products created by other OEMs that need to interoperate with the *turbo-super-duper* hardware in your network will be unable to do so, because they don't share the same "secret" design specs that allow this. Another way of explaining this is that some extended or nonstandard operating modes are often only achievable when pairing with or using equipment from the same vendor. As a result, you're forced to use only one OEM's equipment throughout your wireless network— you'll need the turbo-super-duper wireless client adapter made by ACME Corp. to go with the turbo-super-duper wireless access point (WAP), also made by ACME Corp.

Data Throughput

A properly executed site survey (discussed shortly) can help you plan against issues that affect the overall throughputs on a wireless network. Poor data throughputs, in user terms, is usually described as a "slow network." Proper wireless network planning and designing will take into consideration the applications, data, and other uses of a wireless network. You can then use this information to design a wireless network that will provide the best throughputs to support various use cases.

The planning and design might also determine, for example, that some users, groups, or departments need to be segmented from others. They can also determine that certain areas need more stringent security requirements than others, or perhaps help you realize that a wired solution will better serve the users' needs in certain parts of the network.

Network Reliability

Network reliability is another important metric by which users measure a network. You can be sure that the users or clients of the network will let you (and upper management) know if the network is flaky and unreliable. Unreliable and erratically behaving networks are not only an annoyance, but they impact user productivity.

Proper planning before deploying a wireless network can help you mitigate issues of unreliability and other unpredictable behaviors on the network. For example, it can be easy to overlook the effects of the locations or positions of physical objects on wireless network—such as furniture, trees, walls, plumbing, and so on. For this reason, the results of a wireless site survey of an empty office space will likely be different from a site survey of the same space inhabited with cubicles and furniture.

In shared spaces or buildings with neighbors, wireless equipment can also affect the reliability of a wireless network. Even though you may not have any control over the operation of radio equipment of other offices, you can at least plan for the optimization of equipment under your control. A site survey can help you spot such issues ahead of the actual deployment of the wireless network.

Estimating Hardware Requirements

It's important that you neither underestimate or overestimate the hardware requirements for your wireless network.

Overestimating hardware requirements can mean budgeting or dedicating too many resources to meet simple needs—recommending an enterprise wireless LAN controller with a 10 GB/s backplane for a two-man bookkeeping business, for example, would be considered overkill. Deploying numerous access points (APs) in a small space when it makes more sense to use a single, well-positioned AP is another example.

On the other hand, by *underestimating* hardware requirements, you may save a few dollars in hardware expenditures today, but you may find yourself wishing you'd spent more to get more appropriate equipment further down the road. Although saving money makes upper management happy, it is more important that you choose the right hardware to meet the needs of the wireless network.

For example, planning a network for ten users and purchasing equipment that is designed to serve exactly ten users should be considered cutting things too close. You should instead plan for infrastructure equipment that would easily scale to at least 20 users. Another example: You wouldn't use a single WAP to provide coverage for a large area when it is makes more sense to deploy multiple APs or repeaters to extend the network's reach.

NOTE   There are no equations or formulas to help you arrive at the perfect network plan; it's mostly about common sense.

Site Survey

A good site survey will yield vital information that will aid in designing and planning a successful wireless network. Site surveys can help in determining cell coverage, capacity planning, network documentation, benchmarking, and accessing user requirements. All these activities impact the final wireless network deployment directly or indirectly.

NOTE   *Cell coverage* refers to the useful area that a WAP's radio waves can cover. Cell coverage is useful information when you're determining the best locations for wireless equipment to meet users' needs.

The site survey generally has two facets. The first is nontechnical and involves fact-finding and paper-pushing processes. It also involves accessing actual user requirements.

The second aspect of a site survey is more technical and physical in nature. It can involve taking measurements, assessing the radio propagation properties of the physical environment, and other similar requirements.

Site Survey Metrics

The technical aspects of a proper site survey involves walking around the site and collecting and recording radio frequency (RF) and data metrics at different points in the site—such as the received signal strength (RSS), data rates, and signal-to-noise ratio (SNR).

The RSS metric affects the overall coverage in the site. The lower the value of the RSS, the farther the network will reach. A -65dBm (decibels below 1 milliwatt) is commonly regarded as a decent value for the RSS at the edges of a wireless network.

The SNR measures the difference between the useful RF signals and the undesirable (and inevitable) background noise. The SNR directly affects the data rates achievable on the wireless network. In a good SNR, more of the useful signal is available relative to the noise signal. So a higher SNR implies that effects of the noise signals are less noticeable. In real-world applications, an SNR value of 35dB (decibels) is better than a SNR value of, say, 10dB.

Network data rates reflect the actual data throughputs achievable on the network.

About RF Interference

You'll spend a good chunk of time gathering different metrics during your site survey. Many of these metrics are used eventually in trying to reduce interference to a bare minimum and in determining optimal network configurations.

Every wireless network will encounter many sources of interference, which can be cordless phones, neighboring wireless networks, fluorescent lights, and microwave ovens, for example. Their effects range from the benign to major. Regardless of the source of the interference, it's an undesirable factor in a wireless network deployment.

RF interference in a wireless network has many side effects. Slow network performance is one example. Because the transmission medium in wireless networks is a shared medium, there needs to be a way to police the medium so that all the wireless clients get their fair share of access. Radio interference can disrupt this policing mechanism and slow down legitimate traffic on the network. The source of the interference effectively hogs the medium, which prevents other devices from using it in a timely fashion. This is one of the causes of what users perceive as a slow wireless network.

It is also possible that, by some freak chance, excessive RF noise can lead to corruption of the data being transmitted over a wireless network. This corruption means that the built-in checks that wireless STAs use to confirm receipt of packets will fail, and the sending STA may be forced to resend the data several times before the receiving STA finally acknowledges it (provided it isn't corrupted). This inevitably leads to unnecessary retransmissions, which can slow down the network.

Site Survey Tools

It is sometimes wrongly assumed that a site survey can be performed only with certain specialized equipment and by special individuals and firms. This assumption is also one of the chief reasons why some people don't bother conducting surveys.

It is true that expensive and specialized equipment is available for performing site surveys, and this equipment will do an excellent job. For example, a spectrum analyzer or an oscilloscope can cost $40,000 and is used to measure the frequency, bandwidth, noise, and distortion characteristics of RF circuitry.

But if you don't own your own personal spectrum analyzer, you can choose from among many other tools that can be used to perform a site survey on the cheap.

With a dose of common sense and commodity hardware and software, you can perform a site survey that will not break the bank. A side benefit of this low-budget approach is that you can use the same (or similar) hardware that will be used in the final network deployment.

Following are some of the tools you'll need to perform a basic site survey for an IEEE 802.11–based WLAN:

- Laptop or other portable 802.11-capable device
- A decent wireless adapter for the laptop or portable device
- Some kind of software to show display wireless RSS on the laptop (however, if your specific software or utility does not show the RSS value, see the sidebar "RSS Alternatives")
- An AP or two
- Two to three different antenna types that can attach to the AP and/or wireless adapter on the laptop
- A tape measure or a laser-based measuring tool
- Site map or floor plan—make a sketch of the floor plan if you need one
- A ladder

RSS Alternatives

Any software utility that helps you manage the wireless properties of your wireless interface will have some way of displaying and expressing the signal strength. This information is often displayed using a number of colored bars of different heights—referred to as a "signal strength meter."

Another common method is to express the quality of the wireless signal as a percentage from 0 to 100 percent.

Let's map our fancy RSS values to five little bars or their equivalent percentages when the actual RSS values are not available. We'll use the following conventions:

- When the wireless signal meter shows all five colored bars, we have an excellent signal (or 100 percent signal strength) at the receiving station.
- When the wireless signal meter shows only three colored bars, we have an average quality signal of around 50 percent or higher.
- We will map three bars or higher to RSS values greater than or equal to -65dBm.
- Whenever the wireless signal meter shows only one colored bar, we have a very low or poor signal.

Finally, when the wireless signal meter shows no colored bar, we have no signal at all. At this point, you can replace all references to "-65dBs" when conducting a real site survey with three bars or more. If your utility uses percentages, substitute "-65dBs" with a signal strength of 50 percent or higher.

The Actual Site Survey

The following procedure is admittedly a bit crude, but it will suffice in getting the job done.

One of the objectives of this sample site survey is to determine an optimal location for the APs in the given space. For our purposes, the optimal location within the space provides the best signal reception to the wireless clients while avoiding sources of interference.

Our second objective is to determine the number of APs that will provide good coverage for our site.

First, let's determine the optimal locations for the APs:

1. Power on the laptop.

2. If the laptop or other device does not have a built-in wireless adapter, install the wireless adapter and any necessary software (such as drivers).

3. Attach the external antenna, if one exists.

4. Grab the floor plan for the space to be surveyed, or create a quick sketch of the floor plan if one does not already exist. A sample floor plan is shown in Figure 16-1.

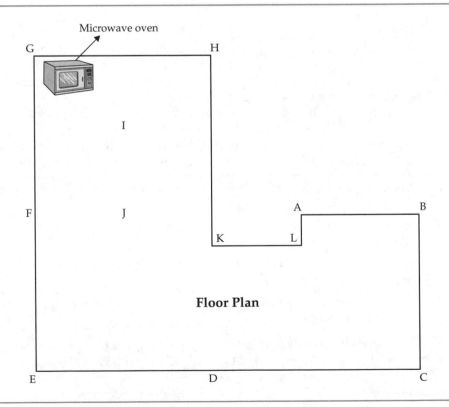

Figure 16-1.   Example floor plan

5. Locate the approximate center of the entire space in the floor plan, and walk over to that location. This is location D on our sample floor plan in Figure 16-1.

6. Power on and place an AP at that location. Make sure that all the wireless security settings of the AP are disabled and that the radio is turned on.

7. Configure the laptop to associate with the WAP.

8. Because of the shape of our sample space, we'll use a patch antenna at location D. Attach the patch antenna to the first AP.

9. Starting at location D, we keep an eye on the wireless signal strength indicator of the laptop and walk back and forth with the laptop in the directions of A, B, C, E, F, G, H, I, and J. We stop walking in each direction after the signal strength drops down to -65dBm.

10. During our sample survey, the -65dBm spot maps perfectly to the following locations on the floor plan: A, B, C, E, and J.

11. Let's rename these exact spots (A, B, C, E, and J) to D_end_A, D_end_B, D_end_C, D_end_E, and D_end_J, respectively (see Figure 16-2). These locations denote the outer boundaries of the RF coverage provided by the first AP. This entire coverage area will be denoted as Cell_A.

12. We'll also rename the location D to D_start.

13. D_start will be the proposed final location of the first AP.

14. These changes are reflected in Figure 16-2.

In some locations on our floor plan, there is still no RF reception. These locations are denoted by F, G, H, and I in Figure 16-2. To find the optimal location for our second AP, we start at one of the outer boundaries of Cell_A.

Specifically, we will start at location D_end_J (Figure 16-2).

1. With the laptop still in tow, walk to the location D_end_J and then walk to the approximate center of F, G, H, D_end_J, and K. This center location is denoted by I in Figure 16-2.

2. Power on and place the second AP in location I. Make sure that all the wireless security settings of the AP are disabled and that the radio is turned on. Configure the laptop to associate with the second WAP.

3. Because of the shape of our sample space (from the perspective of location I), we'll use an omnidirectional antenna at location I. Attach the omnidirectional antenna to the second AP.

4. Starting at location I, keep an eye on the wireless signal strength indicator of the laptop and walk back and forth with the laptop in the directions of E, F, G, H, D_end_J, and K. Stop walking in those directions when the signal strength drops down to -65dBm.

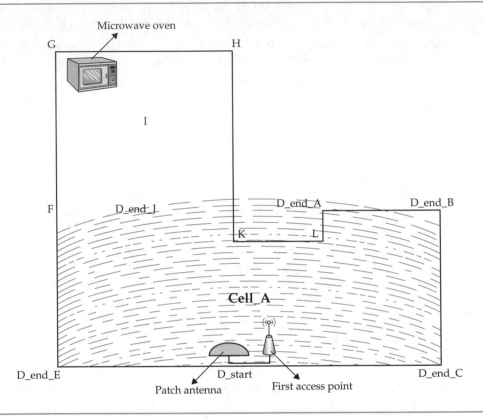

Figure 16-2. First AP location

5. During our sample survey, the -65dBm spot maps perfectly to locations E, F, G, H, and K on the original floor plan. Let's rename locations F, G, H, and K to I_end_F, I_end_G, I_end_H, and I_end_K, respectively. These locations will denote the outer boundaries of the RF coverage provided by the second AP. This entire coverage area will be denoted as Cell_B (Figure 16-3).

6. We'll also rename location I to I_start, which will be the proposed final location of the second AP. These changes are also reflected in Figure 16-3.

And that's it. We now have RF coverage at all locations of our original floor plan. We were able to do this using only two APs.

The information gathered during this procedure can now be incorporated into the planning and designing processes.

NOTE We could have probably gotten away with locating our second AP in location G of the floor plan in Figure 16-1, with a directional antenna at that location. We didn't use location G, however, because of the industrial microwave oven located nearby.

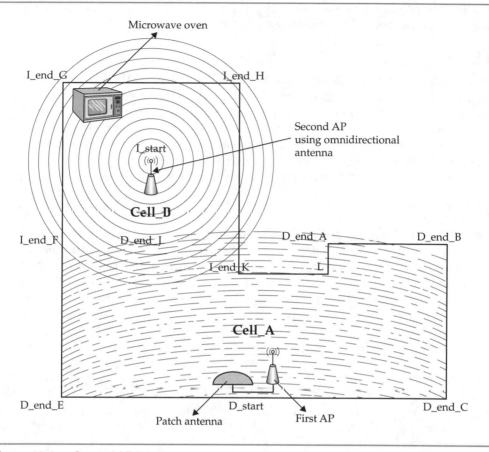

Figure 16-3. Second AP location

  NOTE   Admittedly, the preceding site survey procedure was a bit tedious to follow because the reader has to repeatedly refer back to the floor plan for cross referencing. But although the site survey methodology was a bit crude, you should have gotten the general point of the whole exercise. Also, it should be noted that other site survey experts might be able to find more optimal positions for the access points.

Wireless Site Deployment Examples and Considerations

When you're designing and deploying wireless networks—or any type of networks—you should keep a few things in mind. Beyond the common underlying wireless technology, wireless networks at different sites may have different characteristics, and these differences can be a result of the application for which the wireless network is being used. Some examples are discussed in the following sections.

Hospitals

Hospital environments have numerous RF-based devices in place, such as sensor and alert devices, which serve as lifesaving technologies. Their proper operation can be a matter of life or death. Deploying wireless networks for data communications in hospital environments can be a complicated endeavor. Wireless communications equipment must not cause any interference with other medical equipment. The wireless network administrator may have to go as far as ensuring that certain areas or rooms in the hospital are completely shielded from RF waves from wireless networking equipment.

Airports

Airports offer a mixed bag of sometimes conflicting requirements when it comes to wireless data communications deployments. On the one hand, an airport relies heavily on radio-based communication equipment to communicate between aircrafts and the airport control center (tower). And on the other hand, the airport also caters to the needs of the numerous passengers waiting or in-transit, by providing them a way to access the Internet. The most convenient and practical way to provide passengers access to the Internet is unarguably wirelessly.

As a result of these conflicting requirements in an airport environment, careful network segmentation techniques may have to be employed. This segmentation may have physical and logical components. The physical component has to do with minimizing radio interference between flight communication equipment and wireless data communications equipment. The logical segmentation component has to do with separating the public (passengers) network data traffic from the airport's private data communications traffic.

Because an airport has no easy way of predefining or prequalifying the type of wireless hardware that its customers intend to use on its Wi-Fi network, network planners may need to make compromises and support legacy protocols and legacy wireless technologies to accommodate as many people as possible.

Some airports overcome some of these issues by contracting out the provisioning and management of their public-facing wireless data networks to third parties that use captive portals to provision the wireless clients (passengers). The provisioning function of such captive portals takes care of authentication, authorization, and accounting needs of the clients.

Retail Environments

Retail environments rely on a multitude of wireless communications technologies. Since the dependence on these technologies is usually unavoidable in retail environments, the wireless network administrator must take steps to make the best of a bad situation.

Retail environments rely heavily on wireless bar code scanners, RFID readers, cordless payment processing systems, wireless video surveillance cameras, and demo wireless electronics that are for sale. Most of these gadgets operate in the same unlicensed radio frequency bands used by wireless networking equipment. This all makes for an excellent recipe for interference and general chaos.

A specific use case that is more troubling in retail environments regards point-of-sale systems or payment processing systems that use airwaves that haven't been properly secured. It can be easy for the bad guys to capture interesting network traffic that contains customers' financial data.

Data throughputs and other bandwidth issues are not too much of a concern in retail environments because of the nature of the data being sent back and forth. Handheld scanners, for example, typically transmit only short bursts of data intermittently and do not occupy the RF spectrum for long periods.

High Security Areas

Examples of high security areas are banking and other financial institutions, government agencies, and private institutions that deal in sensitive areas (energy, health care, transportation, or defense, for example).

Let's start off with a simple advisory list:

- The Wired Equivalent Privacy (WEP) is broken.
- The WPA is compromised.
- The WPA2 will someday be compromised.
- WPA3, WPA4, …WPA-unbreakable-edition-version-100 will all be compromised sooner or later after they have been developed and invented.

Here's the point: Use wireless technologies for data communications in high security applications only when you have absolutely no choice.

The fabric (radio waves) of wireless communication makes it difficult to secure perfectly. Further compounding matters is the fact that as newer and stronger encryption methods emerge, the processing powers of commodity computers also keeps increasing, which helps in detecting and exploiting the weaknesses in the methods.

If a wireless network is absolutely necessary in a high-security environment, you need to put some best practices in place. These types of environments might need multilayer protective mechanisms and a distributed infrastructure. This can be expensive, because it means getting a dedicated server for all the individual components that could have otherwise been consolidated. But a distributed security infrastructure means that a compromise or weakness in one of the components does not automatically mean a compromise of the remaining components.

Home Environments

Home users should adhere to all commonsense wireless security principles and properly protect their home wireless networks.

The following scenario shows what can happen if you don't properly secure your network:

1. A home user, Mr. A, subscribes to an Internet service provider (ISP) to access the Internet. No problem.

2. The ISP has some of Mr. A's personal information, such as his credit card information, address, phone number, and so on. No problem.

3. Mr. A is connected to the Internet and all is going well. No problem.

4. Mr. A decides that he needs a wireless network at home, so he buys all the necessary hardware (WAP, wireless adapter, and so on). No problem.

5. Mr. A gets back home and quickly hooks up the new wireless hardware to his ISP's network without securing the network. Problem!

6. Next door lives Mr. B. His Internet service was discontinued years ago by another ISP because he violated the ISP's terms of use. His violations included illegal downloads and attempting to buy merchandise online with stolen credit cards. He did his time on probation and is now back to his usual tricks with another ISP.

7. Using his laptop, Mr. B easily connects to Mr. A's unsecured wireless network to use Mr. A's access to the Internet. Mr. B resumes his illegal and illicit activities on the Internet.

8. Mr. A's ISP is notified that a customer is doing untoward things on the Internet. The ISP investigates and finds out that the offending activity is originating from the equipment registered to Mr. A.

9. The ISP alerts the police, gives them Mr. A's address and phone number, and then discontinues providing service to Mr. A. Mr. B gets off free and clear.

10. The police visit Mr. A, and the evidence against him is very strong. Mr. A may be held liable and accountable for any crimes or illegal activities that malicious third-parties (aka Mr. B) conducted against external networks while using Mr. A's home wireless network.

11. Mr. A gets himself a good lawyer. Fortunately, the lawyer has read this book, so he understands how things like this can happen. The lawyer is able to prove Mr. A's innocence, but Mr. B is still a free man.

12. Mr. A has learned his lesson and secures his home wireless network.

Summary

The Quad-W rule for deploying wireless networks is simple. You learned how it's used to survey, plan, and design a wireless network. You learned how to conduct a site survey on a budget by walking through an actual site survey procedure using commodity hardware. You also learned about the importance of interference and security in several types of wireless network deployments.

APPENDIX | Troubleshooting Wireless Networks

Key Skills and Concepts

■ Learn basic and general troubleshooting techniques.

■ Learn how to troubleshoot specific wireless networking issues.

Throughout this book, you've learned how to assemble and build all the pieces of a wireless network in a controlled fashion. You learned about preselecting the infrastructure hardware, choosing the client-side hardware, selecting the security mechanisms, selecting the operating systems (OSs) that would be used on the client systems, and so on. After such careful planning and staging, any wireless network should just work seamlessly.

In reality, this is not always (or often) the case. You might not always have the luxury of designing and building wireless networks from scratch. And even if you do, things may not always work out as perfectly as they did for us in our examples. A common scenario is that the wireless network administrator will inherit or manage existing wireless networks. Another scenario is that you might be responsible for building and connecting a wireless with an existing wired network. And, of course, existing networks come with their own set of issues or constraints within which you will have to operate.

This appendix is dedicated to scenarios for which things just don't work out like they should. The troubleshooting methodology will work like this:

■ We will go all the way back to the beginning of this book and work our way through each chapter and topic.

■ Under each chapter and topic, some of the things that can go wrong will be discussed, as well as the possible fixes.

■ Some topics (and chapters) will have more troubleshooting information than others, while some topics may not have any relevant troubleshooting aspects.

 TIP   Even though this appendix has been organized into different sections based on some of the preceding chapters, issues can arise from a single or a combination of any of these areas. So, in general, you should always aim for the lowest hanging fruit when troubleshooting. This means that you should address the issues that are easiest to solve or the issues that have the least impact on the network before bringing out the big guns. The ability to easily recognize or identify the low-hanging fruits comes with experience.

Let's begin with troubleshooting regulatory and technical organization issues.

Troubleshooting Regulatory and Technical Organization Issues

The reasons for misbehaving or improperly functioning wireless networks that can be attributed to regulatory issues can be simultaneously salient and trivial to troubleshoot.

As long as you remember that various countries and regions have different radio spectrum usage and allocation rules, you'll be ahead of the game.

So if, for example, you purchase infrastructure-side wireless hardware that was originally designed to be used in Asia, Australia, and the Pacific Rim (Region 3), you should not expect it to work properly with client-side hardware that was designed to be used in North and South America and the Pacific (Region 2). Besides the obvious interoperability issues, you may also be breaking local radio spectrum usage laws by operating wireless equipment that does not conform to local rules and regulations.

In a similar vein, hardware used for wireless wide area network (WWAN) communications or mobile broadband communications are also similarly affected. So, for example, don't expect to be able to set up a mobile broadband adapter that uses Global System for Mobile (GSM) in a country or region where Code Division Multiple Access (CDMA) is the prevailing cellular technology. If, however, your hardware supports multiple modes or bands of operation, you stand a better chance of being able use the hardware in different regions of the world.

Troubleshooting Client-Side Hardware Issues

On the client side, problems that can be attributed to hardware are pretty straightforward to diagnose. Having said this, it should be noted that software issues can sometimes manifest themselves in hardware, thereby making the hardware appear faulty. But once you have ascertained that everything is fine on the software side, you can proceed to diagnose client hardware issues. The subsequent sections cover software troubleshooting methods.

Classic technology troubleshooting techniques start with a systematic elimination and substitution process—you swap out suspect or dodgy hardware with other hardware that is known to be good. Because typical client-side hardware has become so commonplace and affordable these days, it may be prudent and expedient to begin your troubleshooting efforts in this way. Again it's all about the lowest hanging fruit—if it's quicker and easier to change the hardware, and your intuition tells you the hardware is the problem, you can start from there.

Another cost-effective and quick approach to debugging and fixing hardware-related problems is to look out for hardware switches and/or buttons that may need to toggled to enable or power-on the wireless component in the device.

Troubleshooting Infrastructure-Side Hardware Issues

Unlike the client-side wireless hardware, infrastructure-side equipment is often more expensive and as such may not be so easy to swap out. In addition, infrastructure devices these days are often multifunctional, in the sense that the device may serve multiple roles or functions on the network. It may, for example, be used by both wireless clients and wired clients. So if the device needs to be taken off-line for whatever reason, other critical network services that are not wireless related may also be impacted.

Use the following information to help narrow down and troubleshoot problems that are caused by infrastructure side hardware:

Symptom   **Absolutely nothing is working, but everything seems fine at the client end.**

Possible Fix   Reboot or power cycle the infrastructure device.

Impact   All network services that depend on the device might be affected.

Mitigation   The network should be designed using a layered approach. This means, for example, that outages that affect wireless devices should not affect wired devices, if separate layers of hardware are used for the wireless and wired devices. This will help prevent having a single point of failure.

When financially feasible, equipment spares should always be handy. The spares should, of course, be configured ahead of time. Having spares handy will help when infrastructure hardware must be sent off for repairs.

Symptom   **Only certain aspects of the network service provided by the infrastructure hardware are functioning.**

Possible Fixes   If available, review the system or error logs that the device generates to understand the problem.

When possible, restart the specific service or functionality provided by the equipment.

Reboot or power cycle the equipment when all else fails. If the problem persists, contact the hardware manufacturer for support.

Impact   Certain network services or functionality may be temporarily impacted.

Mitigation   A decent familiarity with your infrastructure devices is useful and will help you know where to look for certain things, such as systems logs; it will also help to know what commands to run or which buttons to click to control different functions.

Because nobody is expected to memorize the commands and syntax for the gazillions of network equipment available today, you should at least know where to look for information when its needed or know whom to ask.

Troubleshooting Issues with the Wireless Building Blocks

As mentioned in Chapter 2, wireless communication technologies rely on some fundamental principles and concepts—the building blocks of all wireless technologies. These principles govern the behavior of radio waves and the propagation of the radio waves as they travel through space, and the building blocks are what make actual wireless communications possible.

Numerous avoidable and unavoidable environmental and physical factors can affect the proper functioning of wireless networks.

In general, having a decent understanding of wireless building blocks can help you in troubleshooting. Use the following information to help narrow down and troubleshoot problems that are caused by building blocks:

Symptom   There is no signal.

Possible Causes

a. The signal source may be switched off.

b. The receiver on the client STA may be switched off or disabled.

c. The wireless client STA may be too far away from the signal source or the access point (AP).

d. Too many physical obstacles or barriers lie between the transmitter and the receiver.

Fixes

a. Verify that the signal source is enabled.

b. Verify that wireless hardware on the receiver is enabled and working.

c. Relocate the STA to a location physically closer to the signal source.

d. Remove the obstacles or barriers if possible.

Symptom   Signal strength is low.

Possible Causes

a. The wireless client STA is too far away from the signal source or the AP.

b. Too many obstacles lie between the transmitter and the receiver.

Fixes

a. Relocate the STA to a location physically closer to the signal source. If possible, use a better antenna on the client STA. If possible, use a better antenna on the infrastructure device to focus or channel the wireless signal.

b. Remove the obstacles or barriers if possible.

Symptom   Strong signal, but no connection.

Possible Cause   Other aspects of the wireless network are the cause.

Fixes   See the troubleshooting infrastructure and TCP/IP troubleshooting issues techniques.

Symptom   Low data rates.

Possible Causes

  a. RF channels are overcrowded.

  b. RF modulation methods are not being used in an optimal way.

  c. The network is poorly designed.

Fixes

  a. Change the channels being used for wireless communications to less crowded RF channels.

  b. Make sure that the transmitting devices as well as the receiving devices are using the most optimal modulation techniques. For example, the use of 802.11b-based clients in 802.11g wireless networks can impact the overall data rates at which the normally faster 802.11g can communicate.

   Remember that the building blocks for wireless communications force a trade-off between data rates and distance. So if you want high data rates, you may have to settle for modulation techniques that provide high data rates at short distances.

  c. Make sure that the wireless network is used for applications for which it is capable of handling. For example, don't expect to be able to stream high-definition (HD) video content over an IEEE 802.11b network infrastructure.

Symptom   Packet loss.

Possible Cause   RF noise or interference.

Fix   Find and eliminate the sources of the interference.

Symptom   Intermittent connectivity issues.

Possible Causes

  a. RF interference.

  b. Oversubscribed wireless network.

Fixes

  a. Find and eliminate the sources of interference. Check to determine whether the intermittent connection happens at certain times during the day. If it does, find out what event coincides with the disturbance and deal with that. For example, you may find that the event occurs mostly during lunch periods when the microwave oven is in constant use.

  b. This issue may be a result of inadequate capacity planning. This means that you may have too many wireless client devices trying to access the network at the same time, thereby overloading the AP.

Symptom   Frequent wireless client disconnections and reconnections.

Possible Causes   The driver for the wireless adapter is faulty, or other parts of the system to which the wireless adapter connects might be faulty.

Fixes   Look to the device manufacturer for updated drivers for the client adapter. If possible, switch out the wireless adapter for a known good one and see if the problem persists. If the issue continues, something is wrong with other hardware parts of the wireless station.

Problems with Establishing Wireless Connections

To troubleshoot issues that can occur when a wireless client STA tries to establish a wireless connection, you need to understand the steps or stages involved during the process: scanning, selection, authentication, association, and TCP/IP configuration. Problems can arise during one or more of these stages, and your being able to identify the exact stage at which the problem occurs can help to simplify the task of troubleshooting.

Scanning

Wireless clients scan to search and locate wireless networks to which they can connect. Scanning occurs via the use of information embedded within special wireless management frames, called *beacons*, sent periodically by infrastructure devices. Beacons are similar to advertisements in wireless networking. The advertisement contains information such as network name (SSID), wireless channel information, supported data rates, and time stamps.
Scanning can be either passive or active.

Passive Scanning

In passive scanning, the wireless STA is placed in a listening mode. It listens for beacons being transmitted by APs. The STA does not need to do anything special when operating in this mode.

Virtually all of today's APs make it possible to suppress the advertisement of network names through a configuration option that's called something like "Prevent SSID broadcast" or "Hide wireless network name." Passive scanning by wireless clients is a little more tricky but not impossible when the network name is hidden or cloaked by the WAP. The wireless client STA needs to explicitly know the name of the wireless network to join it when the network name is not being broadcasted.

When the option to hide the SSID on the WAP is enabled, it is always a good idea to "unhide" the network name temporarily when you're troubleshooting wireless connectivity issues that may occur at the scanning stage.

Active Scanning

In active scanning, the wireless client STA is placed in a probing mode. This means that the STA actively probes or queries the wireless airwaves for information about available wireless networks. The STA sends out a special type of wireless management frame known as a *probe request*. Available infrastructure devices, such as WAPs, respond to probe request frames with "probe response" frames.

Active scanning can be initiated by a wireless STA under several scenarios. In one scenario, the STA is in a new environment and has no prior knowledge of any available wireless networks. The STA begins an active scan of the so-called "any" wireless network name or SSID. The "any" here simply means that the SSID is not specified or has a null value. All WAPs and other such infrastructure devices that are configured to broadcast their availability respond to the client. The client can then select the desired wireless network name to which it will connect.

In another scenario, the STA is within the range of a wireless network or environment in which it was previously configured to participate. The STA begins an active scan for the previously configured network. This is how portable wireless devices appear to remember and connect to wireless networks to which they've connected in the past as long as they are within physical range of the network.

Selection

The selection stage occurs whenever a wireless client STA tries to connect to a wireless network. Note that selection happens at the client end—specifically, it occurs within the connection management software or utility on the STA. And, more important, the act of selecting an AP or wireless network does not guarantee anything. You can *select* any network, but other factors come into play to ensure a successful final connection. Selection is just one of the steps.

It was mentioned earlier that the management beacon frames contain information that advertises information about the wireless network, such as network name (SSID), wireless capabilities, supported data rates, and signal strength. Some of this information comes in handy during the selection stage. Let's look at how these factors are used in the selection stage.

Wireless Network Name

The wireless network name (SSID) is, among other things, information used by the wireless client STA to identify the wireless networks to which it has previously connected. It is also used to identify the wireless networks to which the STA has been preconfigured to connect.

When duplicate wireless network names appear, other factors are considered when choosing an SSID—for example, the SSID from the AP with the strongest signal will be selected for pairing with the wireless STA.

Wireless Capabilities

Wireless capabilities are important factors to consider when a wireless client STA is trying to select an AP or a distribution system with which to connect. The capabilities advertised by APs can include the data rates supported by the AP, the supported security options, the supported radio channels and/or frequencies, and the supported IEEE 802.11 MAC and PHY types.

The STA compares its own capabilities with those advertised and supported by the AP and makes a selection based on what provides the best fit. As a simple example, a wireless client STA that supports and implements the IEEE 802.11g standard will tend to favor an AP that advertises and supports IEEE 802.11g as opposed to another AP that supports only the IEEE 802.11b standard.

Signal Strength

The wireless STA can use the signal strength information of the AP stored in the probe response frame to select an AP with which to connect. The higher the value of the signal strength, the better the affinity of the STA to want to connect to that AP. The signal strength value can also be used as a sort of tie-breaker when all the other selection factors are equal.

Authentication

At the authentication stage in the wireless connection setup, the wireless client STA needs to prove that it is indeed permitted to use or connect to the wireless or network resources. Sometimes the infrastructure-side wireless devices also need to prove or confirm their own identities, too. This is known as *mutual authentication* and is determined by type of security mechanisms in place. You read about different authentication techniques in Chapter 11.

At this stage, the security mechanisms that have been preconfigured on the wireless infrastructure devices come into play. The encryption and decryption ciphers and cryptographic systems are used.

Needless to say, a lot can go wrong here and can lead to hours and hours of troubleshooting. For example, a single mistyped character can throw things off. Following are some problems and possible fixes for several scenarios.

Problem   Characters were mistyped.

Things to Check and Possible Fixes   If possible, remove the security passphrase from the infrastructure device and client device and start off with no security. If possible, change the encryption keys to something simple that will be difficult to mistype. After you've established that a mistyped character is the problem, you can go back and change to more complicated or stronger passphrases.

Problem   Mismatched security settings: The infrastructure devices and client devices must be using the same or a compatible security setting. For example, if the AP is configured to authenticate clients using WPA2, and a wireless STA is trying to use WPA, the authentication stage will not succeed.

Things to Check and Possible Fixes   Make sure that the security types and options configured on the client STA match those that are configured on the infrastructure-side devices.

To eliminate a mismatched security setting being the cause of the authentication problem, completely eliminate the use of encryption on both the infrastructure-side device and the client-side device. If things work using no encryption, you can work your way up starting from low-strength encryption all the way to high-strength encryption settings. This should be done until you find a mutually compatible setting to use.

Problem   Interference: Under normal circumstances, interference is a nuisance in wireless communication systems, because it simply disrupts the communication. Interference can occur at any stage during the connection setup process or even after the connection has been established. The probability of RF interference occurring at the same instance as the authentication stage is slim but not impossible. When the effect of RF interference is too great, it may prevent the authentication process from completing successfully.

Things to Check and Possible Fixes   Try to eliminate or reduce all sources of RF interference in the environment where wireless communication systems are in use. Possible sources of interference are too numerous to list them all here. But some examples include other wireless devices, other non-wireless electrical devices, and unintentional emitters (microwave ovens, consumer electronics, light fixtures, electric motors, power transmission lines, and so on).

Association

The association stage occurs after the authentication stage has been successfully completed. Once the STA associates successfully, it becomes a part of the wireless basic service set (BSS). During association, the minor little details of the wireless relationship are ironed out.

Association begins by the wireless client STA sending an association request to the infrastructure device. The association request frame contains information such as the data rates supported by the client STA, the name of the wireless network (SSID), and other special wireless extensions. The infrastructure device responds to the association request with an association response.

Things rarely go wrong at this stage, but the association stage can fail even after a successful authentication stage. When and if something does go wrong at this stage, the problem may be out of the user's control, because it may be caused by something as low-level as the software driver. The wireless device driver on the client STA may, for example, be advertising or requesting a functionality that the infrastructure device does not support or understand. RF interference and poor signal strength can also hinder association.

TCP/IP Configuration

Wired and wireless networks today are based on the venerable TCP/IP suite. This may well change tomorrow when something better comes along, but this is the way things stand today. This means that any wireless node that wants to communicate with other nodes needs to be able to speak TCP/IP.

In the TCP/IP configuration stage of a wireless connection setup, wireless client STAs are provided their unique TCP/IP addressing configuration information. This is needed by the STA for communicating with other STAs or with other network resources, and it is also used by other network resources for communicating back with the STA.

Various aspects of TCP/IP configuration as it pertains to wireless networking will be discussed in the coming sections of this appendix.

Problems with Infrastructure Services and Protocols: TCP/IP, DNS, and DHCP Issues

Infrastructure services work quietly and invisibly in the background of any wireless network. These services are essential to the proper functioning of any wireless network and actually make the wireless network useful to the end users.

The infrastructure services and protocols discussed here are implemented in software. Thankfully, most of the services are based on well-defined and mature standards, so troubleshooting procedures of services can be based on standard methods and tools. However, the value and expediency that simple experience provides when troubleshooting infrastructure protocol and service outages cannot be overemphasized. Unfortunately, experience is not something that you learn in the classroom or from a book. Experience is that thing—that je ne sais quoi. Experience is what tells the network administrator to check that the network cable is properly plugged into the equipment when the link or status light is off, before firing off the other complicated debugging tools. Thankfully, you begin to build your experience repertoire once you've fixed a problem more than once.

Following are some generic symptoms and solutions that can help you in troubleshooting infrastructure services and protocol issues. Troubleshooting techniques on specific OS platforms will be covered a bit later in the appendix.

Symptom   Network layer connectivity problems are occurring.

Things to Check and Possible Fixes   If available, check the system error logs.

Recent software changes or bugs may have affected the underlying network stack (TCP/IP stack) of the system. Reset the network stack to a known working state if possible.

These issues may also be caused by problems at the other layers. So, for example, a problem that appears at the network layer may be resolved by checking the physical cabling between any of the network components that depend on such cabling.

Symptom   Name-resolution problems are occurring.

Things to Check and Possible Fixes   Check the DNS server and its logs to look for anomalies recorded there.

Check to determine whether the issue affects a single system or the entire WLAN system. If the issue affects only a single system, the problem might be a misconfiguration at the client end.

Test to determine whether the name-resolution issues affect only the LAN systems or the external or Internet residing hosts as well.

If wireless network clients are able to resolve the names of local systems properly using the DNS server, the problem might be with the upstream DNS server to which the internal DNS server forwards requests.

Symptom   **Wireless network clients are unable to obtain network configuration information (DHCP).**

Things to Check and Possible Fixes   Check the DHCP server, and check the logs of the DHCP server. Make sure that the DHCP server service is actually running. On certain platforms, it may be necessary to explicitly authorize the DHCP server for the network it will serve. Check the scope or range of the addresses that the DHCP server is leasing out to clients.

If there are more wireless clients or wireless devices in use than the range the DHCP server is authorized to manage, only some devices will be lucky enough to obtain a DHCP lease from the server. In this case, you will need to increase the DHCP range to accommodate the additional network clients.

Check to determine whether the issue is related to a specific client device or whether it is a general problem that affects all the networked devices. If it affects only a single device, check for the presence of a firewall at the server or client end that might be causing the issue.

Symptom   **Wireless network clients or devices are receiving incorrect network configuration information (DHCP).**

Things to Check and Possible Fixes   There might be a rogue AP or a rogue DHCP server on the network that is giving out incorrect information to the wireless clients. You need to track down the device and disable it.

Troubleshooting Networking Issues on Different OSs

Wireless capabilities can be found in all types of devices, in different shapes and sizes, platforms, and applications. Regardless of how the wireless capabilities are implemented, the underlying principles and technologies are all similar.

Earlier in the book, we used three popular computer OS platforms for demonstrating wireless client configuration techniques: Windows, Apple OS X, and Linux-based Free Open Source Software (FOSS). Let's continue here by discussing how to troubleshoot and fix different wireless issues on these platforms.

NOTE Some of the techniques and tools used here are OS-independent, but whenever this is not possible or applicable, the specific differences are highlighted.

Table A-1 shows the locations for informational system messages and system error messages, also called system logs, on these three platforms. These log resources provide useful information that can aid in troubleshooting various wireless networking issues on different platforms.

Table A-2 shows troubleshooting tools, problems, and symptoms that can occur at OSI reference model layers 3 through 7. Note that most of these troubleshooting tools and utilities are command-line–based, so you will type the command in the appropriate command interpreter on the platform. In FOSS/Linux-based systems, the command interpreter is also called the shell. Sample shell environments are bash, korn, csh, and sh. You can get to the shell by using graphical front-ends gnome-terminal, xterm, or konsole. The GUI front-end, known as xterm, also exists on Apple OS X. On Windows systems, you can run the commands by using the cmd command interpreter.

Platform	Common Name	Location of System Log Files
FOSS/Linux	System logs:	/var/log/messages
	dmesg Used for viewing the ring buffer of the Linux kernel; shows diagnostic messages as various events happen within the kernel, such as detection of new hardware, loading and unloading of kernel modules (drivers), activation or deactivation of network interfaces, diagnostic information from modules for wireless interfaces.	To view the current contents of the kernel's ring buffer, type **dmesg**. To monitor the dmesg output in real time, type **watch -n 1 "dmesg | tail"**. To view wireless connection authentication, association, deauthentication, and disassociation messages as they are happening in real time, type **iw event -f**.
	iw Displays information about wireless devices and for manipulating wireless device configuration.	
Apple OS X	**Console** A repository for all the various log files created by OS X.	Applications | Utilities | Console
MS Windows	**Event viewer** Displays various information about significant events on the computer. Useful for troubleshooting problems and errors. Records system errors and other events.	Start | Control Panel | System and Maintenance | Administrative Tools | Event Viewer

Table A-1. Locations of Informational System Messages and System Error Messages

Symptoms, Problems, & Objectives	Platform	Tool, Fixes & Things to Check
TCP/IP: Verifying and viewing TCP/IP configuration information	MS Windows	**ipconfig** Used for displaying, manipulating various TCP/IP configuration parameters—e.g., to view current TCP/IP settings of the local STA running MS Windows operating system type: `ipconfig  /all` The command will display IP address, subnet mask, DNS server address, the gateway address, hostname of the wireless STA running an MS Windows operating system.
	Apple OS X	**ifconfig** Used for displaying and manipulating TCP/IP configuration information for network interfaces on Macintosh systems
	FOSS/Linux	**ifconfig** Similar to the `ipconfig` command in Windows. **ip** Used for showing and manipulating TCP/IP configuration, routing functions on FOSS-based systems.

Verifying and viewing wireless configuration specific information	FOSS/Linux	**iw** To view the device capabilities for all wireless devices, such as RF band information, type:
		
		`iw list`
		
		To view wireless regulatory domain information for the wireless device, type:
		
		`iw event -f`
		
		To display the wireless station statistic information, such as the amount of tx/rx bytes, the last TX bit rate, type:
		
		`iw dev <interface_name> station dump`

	MS Windows	**netsh** Use wlan context of the netsh utility. Some example usages are listed next. To display the entire collection of wireless device and wireless networks information, type: **net wlan show all** To display a list of the current wireless interfaces on the system, type: **net wlan show interfaces** To display the properties of the wireless adapter drivers on the system, type: **net wlan show drivers** To display a list of wireless networks that are visible on the system, type: **net wlan show networks** To display the current global settings of the wireless LAN, type: **net wlan show settings**
Driver-related problems	MS Windows	Use Windows Device Manager to uninstall the device and its driver. Reboot and then reinstall the driver.
	FOSS/Linux	**modprobe** Adds and removes modules (drivers) from the Linux Kernel—e.g., to remove a module named ath5k forcefully for a wireless device, type: **modprobe -r ath5k** To load the module back again, type: **modprobe ath5k**

Table A-2. Troubleshooting OSI Upper Layer Problems on Different OS Platforms (Continued)

Symptoms, Problems, & Objectives	Platform	Tool, Fixes & Things to Check
General TCP/IP problems	Universal	**ping** Generates and sends ICMP messages that can be used to test TCP/IP settings of a system and test that other hosts are reachable. Can be used to check that the TCP/IP stack installed on the local system is functioning properly at a basic level. If you suspect problems with the local TCP/IP stack of the STA, you can begin troubleshooting by running a simple ping test to probe the loopback interface on the STA. To test, type: `ping 127.0.0.1` If the above simple test fails, this is an indication that something is very wrong with the TCP/IP stack on the local system. You can begin fixing this by using the techniques in the repair sections of this table. Still, on checking the TCP/IP stack of the local STA, you should try pinging the IP address of the STA. This too should succeed. Type: `ping <IP-address-of-local-system>`. If everything checks out within the local system, you might want to check that other systems are reachable on the local network. Type: `ping <IP-address-of-remote-host>` The above test may fail, in certain situations. For example, if there are intervening firewalls in use on the network, the firewalls may prevent or block ICMP packets which the ping utility uses by default. Host-based firewall software running on the local system may also block ICMP packets.

Address Resolution Protocol (ARP)	Universal	**arp** After verifying that basic things are fine with the TCP/IP stack of the local STA, but the STA can still not ping other local systems, you could trying clearing out the local ARP cache for the STA. If you can ping both the loopback address (127.0.0.1) and your IP address, but you cannot ping any other IP addresses, use the arp tool to clear out the ARP cache.
		
		To view the current cache entries, type:
		
		`arp -n`
		
		To delete the entries, type:
		
		`arp -`

TCP/IP: Repairing If you suspect that the software TCP/IP stack of the local STA is corrupted or otherwise confused, you can try to repair it.	MS Windows	**netsh** Use the netsh configuration utility (see Chapter 13) to reset TCP/IP in Windows. This has the same effect as uninstalling and reinstalling the TCP/IP protocol. Type: `netsh int ip reset logfile.txt`

	FOSS/Linux	This may be as simple as stopping and restarting the different scripts that control the wireless network interface. Different Linux-based distributions have different ways for doing this.
		
		On Debian or Ubuntu platforms, type:
		
		`/etc/init.d/network-manager restart`
		or
		`/etc/init.d/network restart`
		or
		`restart network-manager`
		
		On Fedora or RHEL or Centos systems, type:
		
		`restart network`
		or
		`service network restart`
		
		On OpenSUSE or SUSE systems, type:
		
		`rcnetwork restart` *or* `service network restart`

Table A-2.   Troubleshooting OSI Upper Layer Problems on Different OS Platforms (*Continued*)

Symptoms, Problems, & Objectives	Platform	Tool, Fixes & Things to Check
General DHCP-related issues		You should first make sure that the wireless STA is indeed configured to obtain its TCP/IP configuration information via DHCP. It is possible that the STA is configured to use static or manual IP configuration, and that statically configured information may be incorrect.
		Verify that the DHCP client software on the STA is running. A listing of all running processes on the system, as well as the system log files, are a good place to check for this.
		The wireless STA must successfully complete all stages of scanning, selection, authentication, and association before TCP/IP configuration can complete successfully.

DHCP-related issues: The client may have incorrect TCP/IP configuration or is simply not getting any TCP/IP configuration from a working DHCP server. Try manually forcing a DHCP request.	MS Windows	**ipconfig** To force a request for new DHCP configuration, type:
		`ipconfig /renew`
		To release the current DHCP lease, type:
		`ipconfig /release`

	FOSS/Linux	**dhclient** Used for configuring network interfaces with the DHCP.
		To request a new DHCP configuration information or lease, type:
		`dhclient <name_of_interface>`
		To release the current DHCP lease, type:
		`dhclient -r`

	Apple OS X	Disable and then re-enable wireless interface using the Connection Manager applet.

General routing problems: These problems manifest themselves as the inability of the host system to reach some networks and not others. For example, a STA may be able to reach other systems on the same network as itself but is unable to reach systems on other networks.	Universal	**route** Displays and manipulates the routing table. Examining the routing table on a wireless STA will help show how the STA plans on reaching other networks.
		
		To view the route table on an MS Windows system, you would type:
		
		`route print`
		
		To view the route table on a Linux-based system, you would type:
		
		`route -n`

Unable to reach Internet hosts. A wireless STA that is able to communicate only with other local systems but is unable to communicate with hosts on the Internet or that reside on non-directly connected networks; most likely a routing-related problem.	Universal	Use the route utility to view the local routing table. If the STA needs to reach other nondirectly connected networks, there should be an entry for a default route in the routing tables. The default route entry will have a reference to the "0.0.0.0" network. Whenever this entry is absent, it means that the STA will not know where to send packets destined for foreign networks, such as the Internet. The fix for this is to make sure that the STA has a default route entry in its routing table. In MS Windows systems, you would issue this command at a command prompt (single line): `route add 0.0.0.0 mask 0.0.0.0 <IP_ADDRESS_OF_GATEWAY>` In FOSS/Linux-based systems, you would issue this command at a shell prompt (single line): `route add default gw \ <IP_ADDRESS_OF_GATEWAY>`

Table A-2. Troubleshooting OSI Upper Layer Problems on Different OS Platforms (Continued)

Symptoms, Problems, & Objectives	Platform	Tool, Fixes & Things to Check
General name resolution issues: Common name resolution methods are NetBIOS, DNS, database, local text files. The litmus test for identifying a name resolution problem is when a host system is unable to communicate with other systems using their hostnames, but is able to communicate with the systems by other means, such as via IP address.	MS Windows	**nslookup** Simple utility for querying DNS servers interactively. Please note that nslookup is also available on other operating system platforms. For example, to query the locally configured DNS server for the IP address for the example.com domain, you would type: `nslookup example.com` **nbtstat** This is used for displaying current NetBIOS information that the local system knows about; also used for manipulating the NetBIOS name cache.

	FOSS/Linux	**dig** This is a powerful DNS lookup utility found in many Linux-based distributions. Its syntax is simple.
		For example, to query the locally configured DNS server for the IP address for the example.com domain, you would type:
		
		`dig example.com`
		
		If you think that something is wrong with the locally configured DNS server and you want to query using another DNS server:
		
		`dig @<DNS_SERVER_IP> example.com`

Name resolution issues: Unable to reach external hosts or sites using their domain names. A simple test to confirm this scenario is by replacing the friendly domain name of any well known external site with its IP address and then substituting this information for web address for the site. For example, if visiting this web address http://66.249.80.104 using a web browser takes you to Google's home page while http://www.google.com fails, then you know you have this problem.	Universal	If the wireless STA is able to browse Internet sites using only the IP address of the remote host, it can mean either of the following: The DNS server configured for the STA is unreachable or unavailable. The DNS server is reachable and available, but the DNS server is itself unable to resolve external host names on behalf of the wireless STA. This might be a problem with an upstream DNS server that the local DNS server depends on.

Table A-2. Troubleshooting OSI Upper Layer Problems on Different OS Platforms (Continued)

Summary

Some things that can go wrong in wireless networks were reviewed in this chapter. Specifically, the different subject areas covered earlier in this book as well as the common things that can go wrong in those areas were reviewed.

This appendix is by no means an exhaustive look at troubleshooting wireless network systems, and it's impossible to cover every single thing that can possibly go wrong in a wireless network—a separate book would be required to do that, and that book itself would span about 100 different volumes! I didn't feel you'd want to read such a boring book and so I didn't write it!

Index